Dr. Frank Field's Weather Book

Dr. Frank Field's Weather Book

Dr. Frank Field

G. P. Putnam's Sons
New York

Library of Congress Cataloging in Publication Data

Field, Frank, date.
    Dr. Frank Field's Weather book.

    1. Weather—Popular works. 2. Weather forecasting—
Popular works. I. Title. II. Title: Weather book.
I. Title. II. Title: Weather book.
QC981.2.F53 1981   551.6   81-7382
ISBN 0-399-12634-1   AACR2

PRINTED IN THE UNITED STATES OF AMERICA

The author gratefully acknowledges the assistance of Lawrence Galton in the preparation of this manuscript.

Without his assistance, this book would not have been possible.

To my wife Joan who has weathered the past 34 years with a sunny smile through all the television storms.

Contents

1

Forecasts and Forecasters . . . and How They Got That Way

"NOW IT'S TIME for Dr. Frank Field and the weather."

I hear that introduction nightly. In one form or another, I've heard it many thousands of times over the years.

Much more often, it's nothing so straightforward. When he was anchorman on WNBC-TV news in New York, Tom Snyder had his ways of introducing me. The gentlest would be something on the order of: "And now here is Frank Field, the most beloved man in television in New York, to screw up the weather." Then there was the time the introduction to my segment was immediately followed by a barrage of snowballs, one of which nailed me right between the eyes!

Each weeknight that red light on the TV camera goes on in studio 6B at NBC in New York City and the camera focuses on me and my maps and charts. Across town at the same moment at the competing ABC station, my son Storm Field faces his cameras, checks to make sure his microphone is securely affixed to his tie and prepares for the pleasantry or jibe that the anchorman is about to unleash in his introduction to the weather.

Those scenes are repeated in your own city!

The time is 10:00 P.M. or perhaps 11:00 P.M. The late news is coming on. The reports have been gloomy all day. You know there is nothing in the way of good news to which to look forward. But you have to watch and listen because the weather forecast will be coming. You're not really interested in any more television—but, well, you want to know the weather. Maybe it's the weekend coming up!

And duly, the news anchorman or an announcer says in Los Angeles: "And now with the weather, here is Dr. George," or

11

"Here's Pat Sajak." In Detroit it's, "Here's Sonny," or "Here's Malcolm." In Washington, it may be Bob Ryan with the weather and in Chicago it's Jerry Taft. You can substitute the name of your favorite weather person for your city.

Each evening with the accompanying banter the weathermen and weather girls address themselves to millions of viewers enjoying their dinners or later at night, peering at the screen between their toes.

Is TV weather forecasting serious? Of course. But it's show business, too. And how it got that way is an interesting story.

Radio, almost from its beginnings, had weathermen of sorts. The weather information was brief and basic. The quickie reports often were part of news, disc jockey, or informational programming. They were expected, were given, and that was that.

Often the forecasts came from the U.S. Weather Bureau—directly. "And now," an announcer would say, "we take you to the U.S. Weather Bureau," and a voice that might be unpolished, monotonous would come on to recite: "Tomorrow will be partly cloudy with a high of so-and-so." And that was it.

It was shortly after World War II that a tremendous interest developed in weather and weather reports. People were becoming increasingly devoted to boating, skiing, other sports, to flying. They were becoming increasingly mobile and concerned, for example, about what was happening in Florida in the winter months.

When television came along, it wasn't long before there was some realization of the potential of the visual aspects of weather. And at some point, somewhere, some unheralded genius had thoughts that ran about like this: "You know it would be interesting if we had a weatherman—or girl—an entertaining type of person, who could get up and give us the weather report with visuals and take a little of the heaviness out of the drastic news we are always presenting."

Very quickly, the TV weather forecast turned out to be a very saleable commodity. There were always sponsors for it. It always made money.

Who did the forecasts?

The first weathercaster on television was probably an animated lamb. Its name was "Wooly Lamb" and it presented the weather on NBC, my own local station at the time: WNBT-TV. From 1941 to 1948 "Wooly Lamb" presented the weather and commercials to the lucky few who could afford to buy TV sets. Then live newscasts flourished along with television, and the search was on for live entertainers to present the weather.

To begin with, they were certainly not real weather people, not trained meteorologists. They obtained weather information from the nearest weather bureau station. And they came on mainly to entertain. Personality counted; information was secondary. There were pretty girls doing the weather, some in bikinis, some in flowing gowns—at one time in New York City there were Carol Reed, Gloria Okon, Jan Crockett, who played the ukelele, Penny Wright and Jeanne Paar.

Why weather girls?

Well a CBS promotion director said, "Women sell more." An ABC vice-president put it this way: "We feel that women—or ladies—have greater acceptance than men, because, well, the men prefer to look at an attractive-looking person and the women are attracted, too, because of the fashions they wear."

NBC held out. Sort of. They had a former Miss Pennsylvania on weekends and a male—Tex Antoine—who cartooned the nightly weather report during the week. Then there were piano players who sang the weather. There were the "eccentric" weathermen who came on as clowns or magicians. And there were puppeteers who did the weather with puppets and song. In addition, many announcers who turned out to have performing talent became weathermen.

A prime example is Willard Scott who began in Washington on WRC-TV and is now on the "Today" show. Willard, a jovial bear of a man, formerly Bozo the Clown on kiddie shows and the first Ronald McDonald, admits he tries to be corny. He might come on dressed as a groundhog or the Easter Bunny and deliver the weather. In fact when Willard was signed on for the "Today" show he announced that he was the replacement for J. Fred Muggs, the chimp, who many years ago starred on the early morning program.

That's how TV weather programs began and they largely stayed that way into the early 1960s. All over the country, the primary attempt was to get a personality on the air who could attract an audience and present the weather report any way he or she could that would be entertaining. The exception—those parts of the U.S. such as the Southwest where tornados and severe weather could not be treated lightly.

Most of these people would be briefed by the local weather bureau or even by professional meteorologists hired by the station to tell them what to say. Over a period of time, some learned enough to become quite competent.

In the 1960s, however, with weather becoming of particular interest in terms of space shots, and with science generally beginning to come to the fore on television, some stations began to consider having experts do the weather. They wanted people who could command a certain amount of respect from the audience.

The search was on for people with meteorological backgrounds—men who had worked as forecasters or meteorologists with the air force or commercial airlines; others who had bachelor's or master's degrees in meteorology; people who had trained in other disciplines, such as geophysics, who had supplemented good basic background in sciences with knowledge of weather.

It was a trend; it still is; but by no means have all stations employed professional weather people. There is considerable variation across the country. In some cities, there are meteorologists only; in others, a mix of meteorologists and performers; in still others, performers only.

How do you know the real thing? That is—a meteorologist. It's difficult. According to Chester Newton, the president of the American Meteorological Society (AMS), "Meteorology is somewhat peculiar as a profession in that anyone with a smattering of the terms used, who has access to publicly available information from the U.S. Weather Service, and who has a bit of equipment can look and sound like a meteorologist."

Not too long ago it was discovered that a respected "meteorologist" who not only broadcast on radio regularly in New York City, but also prepared the weather maps and data for the prestigious *New York Times*, was not a meteorologist at all. Further, his credentials were falsified with nonexistent college degrees, even a phony Ph.D. He was fired.

Within a few months he was back, again as a "meteorologist," featured on another radio and TV station.

There are charlatans in the guise of weathermen who appear on TV and on radio and claim to have the ability to predict snowstorms or hurricanes months in advance. That's sheer nonsense. But unfortunately, the management of the stations or the networks are taken in. These so-called long-range forecasters come and go with their guesses!

I remember one such forecaster on the air in New York who, when queried by the press about his busted forecast of a storm on a given day, claimed that the storm he had predicted months in advance had indeed formed and was over Chicago that very day. He was either the world's greatest liar or a nut.

How can you tell if your weatherman or weather girl has professional weather qualifications? It's not easy. One way that the American Meteorological Society (AMS) has tried to bring legitimacy to TV forecasting is to offer a seal of approval to those broadcasters who uphold a professional attitude toward weather reporting.

Within the AMS, professional qualification is recognized by electing an applicant to membership. Requirements include a baccalaureate or higher degree in meteorology or a related discipline with application to the advancement of atmospheric or hydrospheric science; an adequate record of service, skill and knowledge in work commensurate with that of a graduate holding such a degree; or a suitable combination of education and experience.

But you should understand that there is no license required for an individual to qualify as a weatherman on TV. With looks and a good voice and a friend who runs a radio or TV station you too can disseminate weather! Those who do include weathermen who paraphrase the weather information furnished by a "weather wire" supplied by the National Weather Service, or telephone briefings from the National Weather Service or a private weather firm, or weather stories and reports supplied by the wire services of the UP and AP.

If you become good enough at it, perhaps read up on the weather or take some college courses, you may then venture out and arrive at your own predictions. There's no law against it. In fact I know one weatherman who makes a darned good living by making short-range and long-range predictions for industry. He's also a television personality but he's had little formal meteorological education to speak of.

On the other hand there are qualified meteorologists, trained in the armed forces or at universities, who maintain their own weather stations and prepare their own forecasts for radio and TV. Their numbers are not that great. It's been estimated that perhaps 25 percent of the TV weather performers fall in this category. A few examples are Bob Ryan of WRC-TV in Washington; Al Duckworth, WWL-TV New Orleans; Roy Leep, WTVT Tampa, Florida; Dallas Raines, on Cable News Network; Terence Kelly, WKOW-TV Madison, Wisconsin; Peter Giddens, KGO San Francisco: I hope the others whom I have omitted will forgive me for not mentioning their professional qualifications.

I got into television weather forecasting in a somewhat roundabout fashion. I graduated from college with a degree in geology. When I went into the air force during World War II, there was little need for rock throwers. However, there was a need for meteorologists and Frank Field, like it or not, was picked to become one. I was sent to Brown Unviersity and then the MIT-Chanute Field Schools of Meteorology and became a weather officer with the rank of 2d Lieutenant, assigned first to Mitchell Field in New York and then to MIT in Boston.

I taught meteorology for a while at the Air Transport Command in Homestead, Florida, and was then transferred to the Eighth Air Force in the European theater where I was a staff weather officer.

16

After leaving the air force in 1947 I went to work as a professional forecaster in the New York City office of the U.S. Weather Bureau. It was here that I first encountered the communications media.

The weather bureau issued regular public forecasts and was responsible for preparing the daily weather maps that appeared in the *New York Times* and other local papers. The forecaster on duty also prepared and wrote scripts for radio broadcasts.

Television was just coming of age and another responsibility was the teaching and briefing of the new breed of weather personalities. Most of those we worked with had little knowledge of weather and were reluctant to learn. For the most part they had been given the assignment by their stations and felt that TV was just a flash in the pan and would never take. They would much rather have spent their time doing commercials on radio which in the 1940s was the popular medium of communication.

I also learned the woes of public weather forecasting. There was the day in December 1947 when the chief forecaster of the U.S. Weather Bureau office at 17 Battery Place called for light snow to end during the morning hours. I had just been assigned to the office and was working under him as a trainee.

"The center of the storm is just off Montauk according to my map," he said, "so we should find the snow tapering off now as the center moves off the coast."

We had no weather radar in those days, so we couldn't have known that the storm had stalled. The snow did not end. It kept coming despite the next two optimistic forecasts which were issued at noon and that evening.

The rest is weather history. The city was paralyzed by its worst storm of the century. The storm finally ended but not before more than 2 feet of snow had fallen. I spent several nights sleeping on a desk in the climatology office. I don't think our chief forecaster slept much at all that week or in the weeks which followed.

Of course the missed forecast was widely held up to ridicule, and my heart went out to my boss. But then my own turn came quickly enough.

17

There was to be a huge ticker-tape parade which began at Battery Park and went up Broadway. I was on the midnight shift and scheduled to make the early morning forecast. My forecast, issued at 5:30 A.M., was for heavy rain to end around 8:00 A.M.

The mayor's office was on the phone almost every fifteen minutes. Was I sure? Would the rain really end? The cars were all open limos! Should the parade be delayed?

With as much authority as a twenty-five-year-old professional forecaster could muster I told the mayor's secretary, "Yes, the rain will be over. Don't worry." I went on to explain that the last of the heavy rain was over western New Jersey and should clear us by 8:00 A.M. to 9:00 A.M.

Well it didn't! The parade was delayed two hours with frantic calls coming in from the mayor's office all through the morning. Up Broadway to City Hall they went, under umbrellas in open cars in the downpour. Cussing me, I'm sure, every foot of the way!

The rain did end. Right in the final moments of the mayor's presentation. And when it was all over the sun came out to dry off the soggy ticker tape that clogged the path of the parade. Three hours too late!

The meteorologist in charge of the office was kind. He told me not to feel too bad. "There are worse things that can happen." He was right. They did. On future forecasts.

You get some strange and ambitious notions as a young man; or, at least I did. I liked weather work but I also had become very interested in medicine and health. So, willy-nilly, I applied to medical, dental and optometric schools.

It had to be nilly when, having been accepted to be a future physician or at least to be trained to become one, I had to face up to reality. There wasn't any way I could keep working at the weather bureau in order to support a wife and a newborn baby and go to medical school at the same time.

So I hit on what I thought would be a happy compromise. Keep working and go for a degree in optometry which I did at Columbia University and Massachusetts College of Optometry.

So now I had an O.D. degree. Almost immediately came an

opportunity I couldn't turn down—to set up my own business in meteorology. I established a meteorological consulting office in Manhattan. We performed statistical research dealing with climate for advertising agencies and business, weather forecasting for private industry, testifying in court and helping law firms prepare for weather-related cases. International Weather Corporation, as the business was called, indirectly led me into my future career in television.

When new weather personalities were hired to perform on television there was a need for training. Most had auditioned for the job with little or no background in meteorology. It was my job to teach them the basics and work with them in the studios, helping to set up the maps and other visual displays. Later on this experience was to pay off.

I soon found myself with another opportunity to earn some extra income. Having learned of my qualifications as a meteorologist, Dr. Leonard Greenburg, the first commissioner of the Department of Air Pollution for the City of New York, hired me as private consultant to the department.

This relationship led to the next step in my checkered career. Dr. Greenburg left the Department of Air Pollution and became chairman of the Department of Preventive Medicine at Albert Einstein College of Medicine. He invited me to accept a full-time faculty position at the medical school—to do research on air pollution and its effects on health.

It was a time when the serious nature of this environmental hazard was just beginning to gain attention. Our studies were varied and resulted in such publications as: "Area Meteorology: A Component of Air Pollution" (*Archives of Environmental Health*); "Air Pollution and Morbidity in New York City" (J.A.M.A.— *Journal of the American Medical Association*); "Report of an Air Pollution Incident" (*Public Health Reports*); "Air Pollution, Influenza, and Mortality in NYC During Jan-Feb 1963" (*Air Pollution Control Association*); "Intermittent Air Pollution Episode in NYC, 1962" (*Public Health Reports*); "Air Pollution Incidents and Morbidity Studies" (*Archives Environmental Health*); "Air Pollution and Asthma" (*Journal of Asthma Research*).

After several years at Albert Einstein College of Medicine, another opportunity arose. It looked like television meteorology was attracted to me or I was to it. In 1958 the NBC station was in need of a Saturday night weatherman and was holding auditions. Since Frank Field had tutored most of the weather personalities on the air, why not try him?

The director for the auditions was a red-haired young man with a sense of humor who managed to allay my trepidations. His name was Roone Arledge. He's now president of ABC News and president of ABC Sports.

It's one thing to teach—another to perform. But Roone managed to guide me through my audition tape and I won the job. Years later when I became the full-time weatherman for WNBC-TV, Roone, who was now at ABC, sent me a telegram: I THOUGHT I TOLD YOU TO GET OUT OF THE BUSINESS. BEST WISHES. ROONE.

During the week, I continued to work at Albert Einstein. Saturday I was a TV performer. Not a good one I must admit when I look back at those early tapes.

Before long, I was on TV on Sunday night, too. And after a time, it was week nights at 11:00 P.M., and then 6:00 P.M. as well, and I was WNBC's full-time weatherman in residence, which meant giving up air-pollution research at Albert Einstein.

Well, not quite full-time weatherman. Once I got into the swing of developing and presenting weather reports, I had some time on my hands. Just about then, television was becoming more interested in reporting science and medicine. So, in addition to being a weatherman, in 1962 I became the first full-time meteorologist-science reporter on TV.

And so my days now are filled with both weather and science— quite full days they are, too, beginning at 8:00 to 9:00 in the morning and going through to 11:30 P.M. I've learned to get along well on four or five hours' sleep a night, with some catnaps thrown in during the day. In fact, I have the happy faculty of being able to doze off briefly at night on the set and wake up in time to do my turns. Of course, TV being what it is, and TV people being the jokesters they are, my compatriots on the show delight now and then in turning a camera on a dozing Field.

The younger Fields don't seem to doze—although, in one way or another, they follow somewhat in their father's footsteps.

Storm Elliot David Field, thirty-two, is, in fact, my direct opposition, even as to time, on WABC-TV in New York as I mentioned earlier.

He attended McGill University and, while acquiring a B.A., took a minor in meteorology. Emerging from McGill, what do you suppose he decided he'd like to do? Right, study optometry, which he did at Massachusetts College of Optometry. He then went to Maine and joined an older man in his practice. That lasted six months.

He was, he said, bored, wanted to be a science reporter, and came back to New York. I said: "Good luck." And off he went and got himself a job as science reporter with a small TV syndication house. From there he went to Channel 11 in New York—and, not long afterward, to WABC-TV. And there he is now, opposing me, pointing to the left at almost exactly the same time I point to the right, and vice versa. Put two TV sets together, one tuned to WNBC and the other to WABC, and you can see Storm and me pointing at each other. I really like it. His mother? She faces a daily decision: whom does she watch?

With Storm accounted for, that leaves Pamela and Allison. Any ideas on what they are doing?

Pamela completed two years of medical school and then decided to join the other Fields in television. She began by writing and researching and now she produces and cohosts our nationally syndicated program called the "Health Field."

Allison, having just completed a master's in science communication at Boston University, appears to be headed in predictable fashion along the same paths as her big brother and sister. She is now a reporter on WNBC-TV and sometimes her stories follow my weather reports with appropriate comments from the anchor team.

And how did other forecasters get that way? I've managed to

worm the information out of several topnotchers across the country.

For many it was chance. Being there at the right time and being prepared. Dr. George Fischbeck was lecturing on TV in Albuquerque when the opportunity arose. The station needed a weatherman and Dr. George had previous air force training. He got the job.

Don Kent is probably the granddaddy of TV-radio weathermen. He has been on WBZ Boston since 1951. Even as a youngster Don always wanted to talk about the weather. He worked without pay for three years at a local radio station, talking about the weather. He even paid his own carfare. But when WBZ needed a weather forecaster Don was ready.

Bob Ryan of WRC-TV Washington was involved with atmospheric research with the Arthur D. Little Co. in Cambridge, Massachusetts. He viewed the daily weather shows on TV and thought he could do as well. So when he spotted an ad for a TV weatherman, he applied for the job and won the audition.

Jerry Taft of WMAQ Chicago is a former air force pilot who holds degrees in meteorology. While with the air force, he supplemented his income by appearing on a local TV station in the Southwest. When he left the service, TV weather became his career.

How does one become a weather performer (as opposed to a forecaster)? Again, the stories are varied. Chance . . . design . . . luck. Some use the opportunity as a wedge into the television business. It may come as a surprise to you to learn that some of your favorite news people or television performers once presented the weather on radio or TV. Tom Snyder in his early TV days did the weather, as did Tom Brokaw. On my own station in New York City, all three of the anchormen—Jack Cafferty, Chuck Scarborough, John Hambrick—at one time experienced the thrill of reporting the weather on a regular basis. But they probably were not temperamentally suited for the job because they moved on. Ed McMahon of the "Tonight" show did weather on WCAU in Philadelphia.

You know, of course, what the TV weather forecast you watch is like. Is it typical?

There really is no "typical" weather report. There are almost endless variations. However, a comprehensive report may begin with a little light banter between weather forecaster and news-show anchorman or anchorwoman—a kind of bridge from news of very serious nature to something at least a little lighter.

Temperature, of course, will be covered—the present reading, high and low for the day. Sometimes there may be references to record temperatures for this date.

Humidity will be indicated—usually expressed as current relative humidity. Relative humidity means that the amount of moisture in air at a particular temperature has saturated it to a greater or lesser extent, making you feel more or less uncomfortable or comfortable.

People want to know about humidity. It tells them whether it is raw in winter, humid in summer, what it's going to be like when they step out of their heated or air-conditioned homes, and how to dress!

Obviously, wind is important to many people—if perhaps unimportant to just as many. Wind information is vital, of course, to sailors and pilots, and to people who live at the shore where a strong wind may pile water up on the beaches, bang around boats at their slips, and cause erosion. But many other people just want to know if it's windy out and, when they see flags whipping around, how windy. So wind direction and velocity are given.

Barometric pressure is a pointer to probable weather changes and it is usually given with an indication of whether it is rising or falling or steady. Vast numbers of people own barometers; they like to walk over and see if their readings are the same as the official readings. (Incidentally, there may be some differences that have nothing to do with accuracy of the home barometer. Forecasters give the barometer reading at sea level. A barometer registering in a hilly area or even high up in a metropolitan apartment house may, correctly, produce a different reading.)

It's not, of course, the barometric pressure reading itself but how the pressure is changing that provides a clue to weather changes.

The pressure is a reflection of the weight of the air on top of us. When it is falling, it indicates that a low is heading our way; when rising, that heavier air and a high is moving in.

Go to sleep at night with the barometer reading 30 inches, let's say, and wake the next morning to find it reading 29 inches and, with that 1 inch drop overnight, you may be tempted to predict, as many people would, that a storm is heading in.

And you could be right. But not necessarily. Other factors enter the picture and the barometric pressure and its shifts up and down provide only one clue. In the northeastern part of the United States, you can get a rising barometer, which (so the time-honored but not necessarily accurate legend on the barometer itself may indicate) should mean fair weather. But if there is a northeast wind to go with the rise, you could easily get freezing rain, wind, sleet and snow.

After the barometric pressure and other readings are given, a radar depiction may follow. Radar can provide an approximation of weather patterns out to as much as a 300-mile radius. It can pick up major thunderstorms, rain areas, edges of hurricanes. So the weather person on TV can point out the patterns and perhaps remark that "As you see, there's a little thunderstorm over here and another over there but it looks like we're not likely to get anything."

From the radar screen, the weather broadcast may move on to a satellite photograph. This may be a still picture obtained through the station's own receiving equipment or from the local office of the National Weather Service, or from an AP or UP wire transmission. Animated satellite pictures—a series of stills put together into a film strip—may also be used. Such satellite photographs provide a wide view of the weather.

Satellite pictures are of course Nature's own weather maps. The earliest satellite photos, beginning in 1960, provided strips of individual pictures because the eye in the sky was polar orbiting.

With each pass around the earth, pictures of the daylight side were transmitted to ground receiving stations.

Six years later on Dec. 7, 1966, the first ATS–1 was placed in a geostationary orbit over the Equator. That is, as the earth turned the satellite kept pace with it, appearing to be locked into position, and offering full disc photos of the earth.

During the early days of U.S. manned space flight I found the use of the new pictures from space to be a blessing. When Frank McGee, Chet Huntley or David Brinkley asked for the weather in the recovery zones, I turned to the latest transmitted weather satellite pictures from space. Millions of TV viewers became aware of this new tool in weather forecasting.

On *Gemini* 9 Hurricane Alma was discovered below the originally planned splashdown area. With the aid of the space photos viewers could follow the decision-making process that NASA used in deciding where to bring down the craft.

Cloud cover pictures received from Nimbus-C were used to build a mosaic about 3 feet square which was placed on the NBC Space Center Map. TV viewers for the first time could see the Florida coast, Cape Cod, the St. Lawrence River, the Great Lakes, the California coastline and all the existing cloud formations.

Working as meteorologist on the space flights was an exciting period for me. It did have its ups and downs though. During an emergency on *Gemini* 8 all three networks went on the air when it was believed that the spacecraft would have to be brought down immediately. I knew there was a Coast Guard weather ship in the recovery zone, and placed a call. *The New York Times* on March 17, 1966, noted, "At 10:19 Dr. Frank Field, NBC's weatherman, established contact with a United States Coast Guard ship in the recovery area. He did it by telephoning a Coast Guard station in San Francisco and asking the station to put him in radio contact with the ship. His conversation was carried on the television network."

Well it worked that time! But when I tried the same stunt on a subsequent flight, it didn't. Here's the item from the *New York*

Daily News: "Hoping for a coup, Frank Field, NBC weatherman, tried to reach Ocean Station Victor in the Pacific on radio telephone via San Francisco. There was difficulty. At one point San Francisco radio was clearly heard saying: 'I don't hear you on the line, Mr. Field, but I'm listening to you on television. You're looking good.'"

Later in the space coverage by NBC I managed to bounce back. During *Apollo 8* in addition to covering liftoff and recovery weather, I resorted to a live telephone contact. From a *New York Daily News* column: "NBC-TV's weatherman Frank Field scored a partial coup when he relayed an exclusive eyewitness report on the descending Apollo capsule, as told by Capt. J. M. Marcum, pilot of Pan American flight 811 flying nonstop from Honolulu to Sydney. The jet was at 35,000 feet and spacecraft was reported nearly dead ahead. At 10:40 A.M. Field repeated the pilot's words in which he described the descending Apollo spacecraft as a five-mile-wide fireball with a tail hundreds of miles long behind it. They were fed to him by Pan American dispatcher Jim Lyons in Honolulu, who in turn made the contact with Capt. Marcum. Field later said it was believed to be the first time anyone saw a spacecraft re-enter. Certainly it was an ingenious reporting stunt. But Field was not quite satisfied. He said on the air: 'We had it all figured out so that we'd get a direct description. Instead somebody forgot to hook up our phone here at NBC in New York.' So instead of getting a direct description, Field relayed the conversation he heard via telephone."

In spite of the fact that satellite pictures are now widely used on most weather broadcasts, there are still many viewers who are confused by the picture they see. Are the white areas the clouds or are the dark areas the clouds? The answer is of course that the white areas or blobs or streaks you see on the satellite pictures are the cloud formations. They are the same pictures you would see if you were in a spaceship circling the earth and looking down at the swirling masses of white clouds against a background of blue ocean and brown earth. Of course in black and white the clouds stand out and it's difficult to detect the land and water masses. That's why most weather satellites have land configurations drawn in.

Once in a while your weathercaster may refer to satellite pictures that appear somewhat different. They may be enhanced images to bring out special features. Or they may be infrared pictures taken by the weather satellite. Such photos are shown in tones of gray to distinguish warm clouds from cooler ones and to detect severe weather cores.

Then, on to the weather map, and this can be any of a variety—highly technical, with all the highs, lows, fronts and isobars or stylized, even cartoonlike—on which the forecaster sketches in lines to indicate air masses and boundaries, or fronts.

I prefer to use a simplified map, based on the detailed technical map but focusing on the two or three dominant weather features of interest to us for the next several days.

In some cities where weather elsewhere is of interest—notably but not exclusively in resort towns—a weather "crawl" may be included in the forecast. "Roll the crawl" is a TV station control room term which simply means that crawling up or down the screen will be indications of temperatures and weather in a dozen or more cities, enabling Aunt Bessie sojourning in Miami and coming from Detroit, say, to look at the reading for Detroit (temperature 15 degrees, snow) and chortle about getting out in time.

If there's time, it's possible to include other things—upper-air winds, the frost line across the country, the doings of the groundhog.

And always, of course, there will be extra time given not only during the regular weather broadcast but to bulletins broadcast repeatedly when severe weather—a hurricane or heavy snow, for example—is approaching.

We'll have a further, more informative look at all of the weather elements, at how TV weather forecasts are put together, the bloopers and other hazards of TV weather forecasting, how accurate the forecasts can be expected to be, and what's going on to improve their accuracy, in later chapters, along with some tips, if you're interested, on doing some forecasting of your own.

2

Confessions of a Weatherman

WHAT'S THE WEATHER going to be like for the first two weeks of this coming January? This question was put to me, not by some ordinary viewer, but by an important executive of the National Broadcasting Company.

I explained that the science of meteorology had not reached the state where long-range forecasts could be made with any degree of accuracy. "The best we can do for you," I said, "is to give a five-day outlook and at this time of the year even that is certainly not to be relied on because of the changing weather patterns."

"Well," said this executive, "how come the Farmer's Almanac— or the calendar I used to get from the Aspirin Company—gives you the entire year in advance?" And he added, "They always seem to be right."

I again explained the difficulties in making long-range predictions, that such forecasts for weeks and months were without scientific merit and in most instances simple guessing or based on statistical averages. You know, I don't think he believed me. He wanted to get a two-week forecast for his vacation, months in advance, and he was determined to get it.

That incident brought up once again a chapter in my early days as a meteorological consultant. After World War II, while attending Columbia University I received a call from an advertising agency. The account executive had called the U.S. Weather Bureau at 17 Battery Place and asked if he could get a long-range weather prediction. He was informed that he could not, that such long-range predictions were not scientifically valid. The executive persisted and was finally told by the forecaster on duty at the time that the weather bureau could not provide him with the kind of information he wanted. They referred him to a fellow who used to work there—Frank Field.

"What I need," he said to me, "is a three-month forecast for the Middle Atlantic States. I want to make up a display almanac to put in cigar stores throughout the area right next to a particular brand of smoking products which would offer the almanac."

I must have gone through the same routine he had heard from the weather bureau about the science of meteorology because he cut me short and said, "Look I know what I want and someone is going to give me the three-month forecast because that's what the sponsor wants."

"OK," I said. "If you insist, I will work up a three-month daily weather almanac and it will cost you $300." That was a lot of money to me in 1949, particularly with a wife who was pregnant and with lots of bills to meet.

He agreed.

That evening I sat on the Long Island Railroad with a legal pad and a volume of New York climatology. It didn't take long. I numbered the spaces on my pad for the days of the months and then proceeded to fill in my almanac with such terms as: falling barometer, blustery, snow flurries, rising temperatures, gusty winds, much colder, clearing skies and other general terms.

The entire three-month calendar worked up in this manner and neatly typed was delivered to the account executive at the agency the next afternoon. The $300 check was in the mail and deposited in my bank the next week. It was my first meteorological consulting fee.

I had all but forgotten my very unscientific long-range prediction when I was reminded of it several months later. The account executive was on the phone again. He was quite pleased. The forecast I had worked up for him for the Middle Atlantic States had been 90 percent accurate. Would I be able to make up another almanac for the same states and one for New England? How much would it cost and when would it be ready?

Well, two areas for three months would be more difficult (I'd need two legal pads). But, yes, I could have the almanacs completed in a week (didn't want to make it look too easy), and the price would be $600. He agreed!

Again I went through the same steps, numbering a space on my pad for each day of the summer months—from 1 through 30 for June, 1 through 31 for July and August.

Once more I filled in the spaces with general weather terms: hot and humid, sultry (great for days in July), scattered thundershowers, pleasant winds (they do occasionally happen in June), heat, changeable skies (you never go wrong with this one) and so on.

Well, the Field almanac was a hit the second time around too. Again a call came from the account executive with a request for another highly accurate almanac for the next season. I must confess I accepted the renewal and might still be producing those dubious long-range predictions if the advertising agency had not lost that tobacco account. My almanac went up in smoke. I never had the courage to try to pass it off on anyone else.

So much for almanacs. You can make up your own! What about, say, July 20, 1985? Would hot and humid with a chance of scattered thundershowers work out? I think so. Or maybe February 20, 1985. How about cold and blustery? Bet that won't be out of the ballpark. It's easy. All it takes is nerve. And a legal pad.

And while I'm confessing to you my early indiscretions, let me tell you there are many things I still don't understand about my chosen profession as a TV meteorologist.

A charming and articulate young lady wrote me not long ago to say that she considered weather to be the phenomenon of overriding importance in our daily lives—and didn't I?—and she went on to ask whether, as a weatherman, I didn't view forecasting as "incorporating the best elements of science, magic, observation, clairvoyance, risk, suffering, ecstasy, communing with nature, and feeling at one with the rest of human kind in surviving the vagaries of the weather."

Not bad. I *am* in awe of the weather and only a little less so of the complexities of forecasting.

I am constantly amazed by people's interest not only in weather but in TV weather forecasts and by the strange relationship between much of the public and TV weather forecasters.

Of course weather is important in our lives. It can dictate the way we dress and, to some extent, the way we live. It may even dictate, some believe, the way we feel.

Virtually everyone is interested in weather—and has something to say about it. It's an easy subject to get into: a conversational icebreaker and, not infrequently in a casual conversation, the only subject of discourse. Let two people meet and chances are that they will greet each other with, "Say, isn't this a miserably hot day!" Or cold day, or rainy day, or rotten snowy day. And they may or may not go on from there.

But why is it—and nobody has ever answered this to my satisfaction—that so many people feel they must wait up until 10:00 P.M. or 11:30 P.M. to get a late weather report? "Can't get to sleep without it," many tell me. They are not going anywhere for the next eight or nine hours, and when they get up next morning they will put on the radio or TV anyhow to get the weather report. Why the insistence on presleep weather information—almost as if it were a soporific!

Another puzzler for me: I often hear from the audience, "Why do you put on such a complicated weather report? Who needs it? Why don't you just tell me whether it's going to rain or snow? Why do I need all the garbage?"

But try eliminating any of the "garbage"! We once stopped noting barometer readings and I got endless complaints—and even threats: "If you don't put the barometer readings back on, I'm going to watch your son!"

Once, not too long ago, one of the other major TV stations in New York City got itself a new director of news who promptly decided that it "didn't make sense" to devote all of three or four minutes of valuable time to a weather report when much of that time could be used for news. "We can tell them," he announced, "that it's going to rain or snow in thirty seconds!" And that's what they did.

Their ratings plummeted disastrously. The news director was very wrong and soon out of a job. The same viewers who may have said, "Who needs all the garbage?" now felt they were being shortchanged.

31

"Well, gee," it appears that they decided, "on other stations they talk about the cold wave in Florida where my uncle is, and I feel much better about freezing in New York when I know he is freezing down there."

All the little things that so many people said they didn't care about suddenly turned out to be things they did care about. They wanted to know that Chicago is "buried under" a 24-inch snowfall. "Thank God it wasn't us," they could say.

They wanted to know how cold it got that day. Sure they knew it was cold—but how cold? They may have been working in well-heated offices but when they got home, for some reason, the fact that it never got above 12 degrees was important to them.

My competitor TV station quickly changed its tune. There were auditions for a new weather personality and the weather segment was reinstated.

And yet another phenomenon perplexes me: the very curious relationship between TV weathermen and their audiences.

TV stations hire analysts to make hundreds of random calls to ask people: Do you know this name? Do you like this individual? More than? Less than? And the analysts come up with ratings.

TV stations love ratings. Analysts like to make them. And almost invariably the ratings show that TV weathermen, wherever they are in the country, are very recognizable creatures. More recognizable than news people. And in most cities a TV weatherman is the number one personality, the most popular.

New York City is no exception. I happened to be Number 1 there, and Walter Cronkite—this was before he retired—was Number 2. Which is very weird but that's the way it works out.

Now, it would be the rare person who would go up to Walter Cronkite and say: "Hi, Walter, how's the Middle East coming along?"

But it's hardly rare that a person marches up to Frank Field and immediately says: "Hey, Frank, what happened to the weather this weekend?"

I have been sitting at a luncheon with the governor of New York State and had people walk right by him and grab my collar and pull my arm and say, "Hey, Frank, what about the weather? Boy, did you foul up? Wow, your son is better than you are."

If you're a TV weatherman and if you're not wary, you can be crossing a street and thinking about something else, only to jump 20 feet in the air as a horn blast from a cab goes off in your ear. You're ready to whirl around and scream at the idiot who did that, but there's the cabbie smiling and saying, "Hi, Frank."

Or you can be dining in a restaurant, and as your hand goes up with a spoon and you're about to shovel a little soup in your mouth, someone suddenly grabs your elbow and spills the soup all over you and says, "Hey, you're the weatherman—Frank Field?"

Or you go into a restaurant for the first time and a waiter goes back to the kitchen and says, "Hey, you know who's up front? The weatherman." And the kitchen empties out and the cooks and busboys come marching by your table, peeking down into your mouth as you try to eat.

Or you are coming through a tollbooth and reach out to give the guy a buck—and he grabs your hand and almost tears it off and says, "Hey, Frank, what's it going to be like today?"

I remember once standing on a street corner, waiting to cross the street, in the midst of a snow and sleet storm, with mush up to my ankles. A well-dressed woman was standing next to me and suddenly she reached out, touched my elbow, and asked, "You're Dr. Frank Field?" I said, "Yes, ma'am." And she said: "What's the weather going to be like today?" I said: "Just terrible," and ran across and I could see her still standing there, thinking to herself, "What a fool question to ask." But that's the kind of reaction a TV weatherperson gets.

There's also a strange love-hate aspect to the relationship between weathermen and the public.

If your broker tells you to buy such and such a stock—"excellent prospects"—and you lose $5,000 on the investment, well, you hate the so-and-so and go to another broker. You half expect that a broker doesn't have all the answers.

That's true, too, for a physician. If your doctor tells you that you're in great shape and you go outside and suddenly have a

33

heart attack, well, you may say, how can you blame a doctor for a heart attack?

But the weatherman? Ever since weather forecasting began to be practical, many people have assumed that a meteorologist should be right. It's his job to make predictions, and they have to be correct; no ifs or buts, no room for error. I learned that early on in my weather forecasting days!

In August 1961 the late Guy Lombardo, the famous orchestra leader, complained to the FCC about the weather forecasts in the New York area. Guy, who produced the outdoor summer shows at Jones Beach, complained in a telegram that weather commentators had predicted the worst weekend of the summer, yet not a drop of rain had fallen—but the attendance at his show had!

He was quickly followed by the director of the Car Wash Institute who represented a $2-million-a-day business. He complained that business that summer was off by 40 percent because of "negative" forecasting, furthermore he said that weathermen should use the term partly sunny rather than partly cloudy.

A *New York Times* reporter drew me into the hassle:

Dr. Frank Field, a weatherman for WNBC-Radio and Television stations, disclosed he had received a letter of complaint from a businessman.

"He makes umbrellas," Dr. Field explained. "He says my forecasts of fair weather this week have been ruining his sales."

The *Asbury Park Press*, meanwhile, featured a story in which Mayor J. Stanley Tunney of Seaside Heights, New Jersey, verbally attacked a New York City weather girl, Jeanne Parr, because she had suggested that viewers stay home with a good book rather than go to the beaches, because of the weather. The ad-lib, claimed the mayor, cost the town at least one thousand dollars in beach receipts alone!

I recall a rather poignant sign that hung in the New York City weather bureau office. It stated that "a weatherman's sins are never forgotten, a doctor's are often buried." That's true! I still have people remind me of a bad forecast made years ago—that they cherish as though it happened yesterday.

Yet, despite their high expectations, people still, on the whole, accept errors in prediction with good grace. The attitude seems to be: The weatherman is a friendly fellow who is trying his best, always making mistakes and having to apologize. But he really is trying even if he fouls up.

There's empathy—and a good thing, too. There is also a lot of ribbing. In my early days on weather, I got a huge amount of it from Johnny Carson. He picked on me constantly. Almost every night for a couple of years, he would open up his monologue with a remark such as, "Well, folks, I found out something about our crack meteorologist—what his job was before he came to work in New York. He came out of the service. *He* was our lookout at Pearl Harbor."

Or, "Well, folks, I finally saw the crack meteorologist doing his work today—up on the roof with a lock of hair from the anchorman, doing a dance."

Or, if we had a bad snowstorm—and no matter that I had predicted it correctly—he would come on and say, "Well, by now everybody knows New York is paralyzed by the twelve-inch snowstorm. Everybody in New York, all the weathermen in New York, predicted it right on the button—except [and the audience, knowing what was coming, would start to howl] our own crack meteorologist, Dr. Frank Field." And he would then go on to say something like, "Frank predicted that there would be a flow of lava from the belly button of the Statue of Liberty."

Or, if we had the hottest day of the year, Johnny would come on with a statement such as, "Well, I saw our crack meteorologist, Dr. Frank Field, today—wandering up Fifth Avenue with an umbrella, and somebody told him it wasn't raining, and he said, 'Oh, it isn't?'"

I remember the first time I appeared on the Carson network program. He told me that it seemed that nobody outside of New York believed there was a Frank Field. Everyone believed that he was making me up, and he wanted me on the show; it would be something of a national event. And he would like to talk with me about the science of weather.

What I didn't know was that they had rigged up a very elaborate snowstorm mechanism, with pipes coming over the curtains.

And so Johnny came on and I came on and he said something to the effect: "Frank, it's wonderful to have you here. I wanted the people to know that you really do exist and that you're not as crazy as we thought you were. By the way, it's now March 28. Do you expect we'll have any more snow in New York?"

"No way," I said.

And as soon as I said it, they let go with a snowstorm in the studio that buried me. And the audience howled.

About six months later, Johnny had me come on his show after we had had two weeks of rain. It had been the worst two weeks we had had for a long time, raining every day.

He had a weather map set up on a tripod and he told me, "OK, Frank, why don't you, as a crack meteorologist, explain why we had two weeks of rain."

Dutifully, I picked up a pointer, went to the map, and started to explain. "Well, we have a huge storm that is out in the Atlantic . . ."

At that point, Ed McMahon, a couple of stagehands, and the orchestra leader, Doc Severinsen, all loaded with buckets of water, came up and let me have it. I was soaked, the audience was in an uproar—and so, later, was the then-president of NBC News who hated it. He thought it poked fun at the news department.

With all the jibes from Carson, my popularity grew. Johnny was a godsend! I don't think I could have made it without him!

I first signed a contract with NBC in January 1959. I guess that makes me one of the longer-surviving meteorologists on television. For the first year my on-air uniform was shirt-sleeves, to convey the idea of a working meteorologist at the office. That lasted until a sponsor came along who felt it was more dignified when one wore a jacket while visiting viewers at home.

I was the weekend meteorologist on WNBC in New York for a number of years. During the week I was on the faculty at the Albert Einstein College of Medicine, involved in air pollution and health research. But then in 1966 NBC's star weatherman found greener pastures at a competing station. It was short notice. I was called in and offered the position. With a fanfare of full-page ads in all the local newspapers the announcement was made, "Dr. Frank Field reports the weather scientifically without the gobbledygook." That was a mistake.

Tex Antoine, the premier weatherman, was loved in New York City. He was the number one TV personality and earned more than any anchorman. There was a deluge of mail and telephone calls. What the hell did Frank Field mean by gobbledygook? After twenty years of Tex, who was this smart-ass to denigrate their idol? The heat was intense and it was Johnny Carson who helped to cool things by picking on the "crack meteorologist."

For example, in kidding the news media Johnny described how the various publications and broadcasts would cover the big story of the flood and Noah's Ark if it took place today. The headline in the *Meteorologist News* he said would read: "Dr. Frank Field predicts fair weather for weekend."

Carson's poking fun at me gave me the opportunity to retaliate. For instance, I took a news photo of a man reaching up to help a woman stranded on the roof of a car in flood waters. The woman had a terrified look on her face and I pasted a picture of Carson's head, with a leering expression, on the man's body. Each time Johnny poked fun at me I would use a faked picture.

I also found a full-size cardboard cutout of Johnny Carson that had been used for NBC promotions. I had it placed on the set, and on occasion ended the broadcast by saying, "That's the weather for tonight," and ramming my wooden pointer through his ears or other parts of his anatomy.

I remember one harrowing yet funny experience. It was the night that Johnny did a walk-on during the weather, broke my pointer and flipped open an umbrella. I had no idea he was on the set until the point of his umbrella was firmly implanted in the seat of my pants. I will never know how I got through that broadcast in one piece.

As you know, Johnny moved to California. But by then I think he had made the "crack meteorologist" a household name. I end this chapter of a TV weatherman's confessions with a confession. I really missed Johnny and his pokes at me!

3

A Day in the Life

It BEGINS sometimes before daybreak. A forecast of heavy snow, for example, means rising suddenly wide awake in the middle of the night and looking out the window for any sign of snowflakes. It means a sleepless night, for when the snow begins, it's back to the studio for special weather warnings.

But even on quiet days the weather takes precedence—even on weekends.

If you're a weather forecaster—or at least if you're this one—the very first thing you do on leaving bed is lunge for the window and take a look at what's doing outside. How did I do? Was last night's eleven o'clock forecast right?

Looks good. But I check further even as I go about the business of shaving, showering and dressing. On goes the radio to pick up the weather bureau's continuous forecast. What they're forecasting at this hour for the rest of the day is right in line with my predictions of the night before.

Fine. I relax as I turn the dial to pick up the news and weather broadcasts of several stations. There's some variation—to be expected because different people are doing the broadcasts and each has his own little touch and even predilections.

I reach my office well before 9:00 A.M., not because I must address myself to the weather, but because of other commitments. It might be a taping session. For several years I was cohost of the syndicated program "Not for Women Only." Now it's another program called "Health Field." In addition there are special half-hour shows such as the monthly "Test Show" broadcasts in New York.

If it's not a special taping session that begins early in the day, it's the usual assignment to cover a health or science story for the evening news.

But no matter what the day holds in store for me, I try to stop by at the National Weather Service office which is on the mezzanine floor at 30 Rockefeller Plaza in the same building as our NBC studios. A quick look at the maps to see that nothing unusual is likely to happen and I'm on my way: Today is a quiet day on the weather scene.

I have a clear picture in my mind of the map I used last night on the broadcast and from what I've heard on the weather bureau forecasts I know things are the same. Nothing is going to happen until that high moves across and that will be sometime on the weekend. So I have clear sailing today and probably for the next two or three days.

There is one little worry on my mind, though. Saturday is Saint Patrick's Day. In New York, St. Pat's generally is not a good weather day. For some reason, there always seem to be snow flurries, cold winds, and kids marching up Fifth Avenue freezing to death.

It looks good for this Saturday and we have already gone out on a limb and said it would be a good day. Still, there's a little nagging concern: Will I win this one?

In the office, I check on what major science conventions are in town, whether there is one—maybe even two—I will cover, and whether I have a film crew available.

Today there is a medical meeting that I'll cover with a crew for an hour or so starting at noon. I'll be doing an interview with the man who will be presenting a newsworthy report on a new development in medical testing. The interview will probably make a three-minute feature on the evening news. With the schedule set, I tackle the mail.

There is always a lot of mail from schoolkids. They want copies of satellite photos, information on tornadoes, pictures of clouds. Some will be doing a project on hurricanes and want information on that subject. Almost always there is at least one letter from a youngster asking how I became a weatherman because he wants to be one.

Kids are increasingly knowledgeable about weather. They know about fronts and highs and lows, often more than do many adults. In many junior high schools, and even grade schools, earth science and meteorology are prime courses.

As a matter of fact, many of the young letter writers would like me to stop simplifying the weather maps. They want more detailed weather. Just the opposite of what their parents want.

I generally pass on an important address for the young people to write to. If you know of a young person who is interested in a career in meteorology pass this on. The American Meteorology Society is located at 45 Beacon, Boston, MA 02108. The society has numerous publications and offers different types of membership ranging from student membership to professional status.

My mail also includes many letters from viewers who have interest in the science and health reports I produce on our newscasts. As science editor I wear a second hat. I was responsible for introducing the Heimlich Maneuver on TV in the mid-seventies. It has now become accepted as the first-aid technique for saving individuals who are choking on food particles or foreign objects trapped in the throat.

I still receive numerous letters from people who have been saved by the "bear hug," or Heimlich, thanking us for promoting this procedure over the years with demonstrations and interviews.

And of course, any reports that deal with a new medication or a medical innovation result in a flood of letters seeking more information. I make it a rule to go through all my mail.

The combination of weatherman and science editor has paid off for me. I can relax and enjoy the weather report, joke with my news team and relate to the audience in a light and personal vein. On the other hand, the science reporting adds some stature to my science background and offers a serious note. So I can enjoy the best of two worlds.

Wherever my science-editing assignments take me, I'm back in the office by 4:00 P.M. or before.

Then it's down to the weather bureau again.

It's time to prepare the evening's weather report. I already have a fairly good notion of what to expect on this uneventful day, weather-wise, with its stationary weather system that I've been watching. So I race through the Teletype summaries and stories to see what's been happening around the country.

The coldest spot in the country today was in North Dakota. The hottest was in southern Florida. I absorb these bits and pieces of information along with sundry items about the heaviest rains, areas of thunderstorms, flooding or other items of interest.

Since my weather report is all ad-lib on the air I must be ready to do as little as a minute or as much as ten minutes, if need be. I must also be prepared to answer any question the anchorman throws at me. When does spring arrive? How cold was it this morning?

Now it's over to the posted weather maps that continually feed into the National Weather Service office. I leaf through the upper-air wind-and-pressure charts and the surface weather maps. I chat with the two forecasters on duty, one of whom is responsible for issuing the weather forecasts for New York City and vicinity. The second provides the weather predictions for the airlines at the city's airports.

I may sit at the weather terminal for a few minutes and request the main computer center to provide me with more detailed weather, which is flashed on the television screen in seconds. Or a national picture of the highest temperatures for all of the nation's cities, perhaps, or a close-up of the Middle Atlantic States and the minimum temperatures for the day.

I quickly note the data for the day recorded in Central Park where the city's observation station is located, the low and high for the day, the amount of rain that's fallen thus far this month, what the sky conditions were for the day. And since our viewers live throughout the tri-state area, I check the surrounding areas too. Did it shower in the Catskills? Was there fog along the Jersey coast?

As soon as I'm satisfied that I am fairly well briefed on the details. I sketch the major weather systems I'm concerned with on my map. The positions of highs, lows, fronts, etc. This data I will place on the studio map.

I look through the afternoon's weather satellite pictures and select the one that I feel is most visually informative. This picture will be duplicated electronically by the NBC graphics department and fed into the studio when I call for it.

Now I look back at my previous day's forecasts including the five-day outlook to see if there is any reason to change any of the elements. For example, perhaps I should raise the projected nighttime temperatures a few degrees because I expect a little more cloudiness and less radiation cooling.

Then I'm ready to go.

It's different, of course, coming in on a Monday after a weekend off—and very much different when storms are brewing or upon us.

Monday starts with a considerable amount of research. First I take a quick look back at the weather maps to see what was going on—in detail—over the weekend and, in terms of predictions I made on Friday, whether things went right or wrong, and why.

Then I take a hard look at all the data available about what's going on right now and what could be happening shortly. I examine upper-air charts and also surface maps to see how storm and weather systems have been moving over the past twenty-four hours. There's also a look at computer charts analyzing millions of bits of information to provide some indication of what patterns—global winds and others—are forming or likely to form in the atmosphere.

In addition, there's a look at predictions made not only by the local weather bureau but also by others within 100 miles, at the airways predictions, at marine forecasts, and a look, too, at discussion among experts who have been working for the past several days.

Then it's time to sit down and begin to draw up my own picture of what is going to happen.

Why should there be any need for me to do that when I might simply use the weather bureau forecasts? Because my forecast may be at variance—sometimes even complete variance—with that from the weather bureau.

Weather forecasting today is far from being all science. Rather, it's a combination of science and art.

In that sense, it is not unlike medicine. When a patient presents a problem, the physician can make use of science, ordering laboratory and other tests that may help in making the diagnosis. But not infrequently, he will, while taking into account the results of the tests, depend as well, sometimes more or even entirely, on his clinical judgment. That judgment is, in effect, a gut feeling based on his overall assessment of the patient and his own longtime experience.

In weather forecasting, a gut feeling can count.

Farmers often make fine weather predictions for the immediate period out of gut feelings. They look at the sky, assess cloud formations, a sunset, wind and wind direction, and come to a conclusion—and it is often accurate for the next twenty-four hours or less.

But farmers may not know that there is a huge old cold air mass sitting not far away and ready to break up, or a big storm sitting out in the Atlantic that is going to make a vast difference in the weather.

Knowing that depends upon science, and we get that knowledge from vast amounts of detailed information from the weather bureau network. We also get the official forecast for the area.

In TV forecasting, we start with the official forecast but it's often essential to have the gut feeling of the broadcaster—his or her idea, from experience and from looks around at the immediate vicinity, of what is likely to happen in terms of the viewing area.

Let a TV forecaster move from one area of the country to another and until he gets to know the new area his forecasts are likely to be off.

For years there have been contests for forecasters, and, considering that they are all using the same basic data, it may seem strange that one or two consistently score highest. The explanation lies in their gut feelings.

The science is fine. If you're a TV forecaster, you use the science as it exists. You use such wonderful technological innovations as satellite pictures and cross sections—virtually slices—of the atmosphere. You use the computer output, a distillation of information collected from masses of data. But if you were to put all the computers in the world together so they functioned as one, you still couldn't keep up with the immense number of variables that are involved on a global basis for 10 miles of atmosphere. There is no way to keep up with them all; weather is that complex.

So you develop your skill, the knowledge you've acquired over the years. I've been forecasting weather in New York City since 1947 and I can often, in effect, "smell" the weather. I would have to be pretty stupid if I couldn't do that after all this time.

For example, there's a shift in the wind and it's possible to figure from the data that the projected high is not going to build up if the wind stays that way. And it is possible to figure that the line between low-lying stratus fog, cool weather and northeasterly winds on the one hand and, on the other, warm sunshine, is likely to fall somewhere down through the metropolitan area. Unless you're careful, you can make a forecast—say, for Memorial Day— of fine warm weather, and people who go to the Catskills will indeed find sunshine, 80 degrees, while others going to Montauk at the end of Long Island are zonked out by bad weather. Yet both are well within 100 miles of New York City.

Or you get the picture from satellite photos and other data that a sizable winter storm is moving your way. There'll be snow all right. The question is how much and where. Where is that line going to be between light or no snow and sizable accumulations? That's where gut feelings come in.

In fact, a lot of feelings enter into TV weather forecasting. What do you do, for example, when you are coming up to a Labor Day weekend and you have a very iffy kind of weather situation? Do you go out and say, well, it looks like a rotten weekend, folks. Or do you say, well, it looks like we could have a pretty nice weekend coming up but there's an outside chance that if the storm gets close enough, it may do such and such.

I happen to prefer the latter tack.

The forecast issued may call for generally fair skies—but a study of some of the computer drawn upper-level wind charts out of the National Oceanic and Atmospheric Administration (NOAA) Center indicates a trough along the east coast as a strong possibility.

To the experienced meteorologist, this may suggest a situation in which the rain over the southeastern states might slip northward along the coast and spoil the forecast. Perhaps with a holiday coming up one might want to predict more cloudiness for the area than was indicated in the official forecast. It's called a hedge!

If there are easy days in TV weather forecasting, there are also tough ones—the days of severe storms when you stick entirely with the weather and have no time for science reporting. For example, when one heavy snowstorm hit the metropolitan area, I was in my office at 5:30 A.M. I went on the air at 6:25 A.M. and every hour after that with updates until early the next morning.

When it's snowing heavily and accumulations are mounting rapidly, people are very much concerned about what's happening and you have to tell them. They want to hear from you repeatedly. You have to tell them, "We now have 6 inches in the city and expect 7 inches more. Roads are closed in this area, don't try to drive here, don't do this, don't do that."

And then at some point that afternoon or evening, you can come on and say, "Well, we now expect the snow to taper off in the next three to four hours; the worst is over." You try to allay fears.

And sometime later, you can go on and say, "Well, we're still getting a few flurries but the sun is shining over eastern Pennsylvania and it has stopped snowing in western New Jersey. You folks over on the eastern end of Long Island have another four hours of snow coming but it will end in the city within the next thirty minutes."

When it comes to advance warnings of oncoming storms—storm watches—some judgment is needed on the part of the TV forecaster.

The last thing you want to do is to alarm people needlessly. On the other hand, you hardly want to let them be caught unaware.

Two kinds of judgment can be involved. Sometimes, the TV forecaster, sensing something serious coming up, may set his own

46

storm watch. He may, for example, go on and do a bulletin early in the afternoon, saying we expect heavy rains to move into the metropolitan area this evening.

If the National Weather Service puts a storm watch into effect, the TV person has no choice; he is bound to announce it even if he disagrees. He may use judgment, however, in interpreting it.

It's not enough to announce the fact of a storm watch or a hurricane watch, period. That certainly would frighten people, leave them confused and not really informed.

Instead, the TV forecaster may go on and say: "The National Weather Service has issued a hurricane watch. The storm is now off Cape Hatteras and it could just as well move out into the Atlantic. If you have a boat, you might do well to take precautions. We'll keep you advised."

Or say the weather bureau is forecasting snow beginning in the afternoon with a possible accumulation of 3 to 4 inches. Again you can do some interpreting: "We expect snow to begin in the afternoon. It should be light for the most part. The storm center will be well south of us. Washington is likely to get hit but we will be on the very edge. We might get a couple of inches if it moves close enough."

When a big storm arrives or is in prospect, it can be a long day—and night—for the TV forecaster. On many occasions, I've been on duty for twenty-four to thirty-six hours consecutively, going on the air with bulletins every half hour.

Do I like the job? Most of the time, decidedly. It's an especially rewarding occupation when you've made a forecast and it has worked out and the day is fine and brisk, and you feel sensational.

But then there are the days when your forecasts go sour. Or even those days when you have to forecast "cloudy, windy, raw" and you say to yourself, "What am I doing in this place? The weather is atrocious; I hate this business."

47

4

Bloopers and Other Hazards

ON A FEBRUARY DAY not long ago, a storm was clearly brewing up for the New York area.

You could see it on the weather map in the National Weather Service office downstairs from NBC in Rockefeller Center. The map showed a line from Ohio through West Virginia, Virginia, North Carolina and Florida. And toothlike projections on the line, pointing east, indicated that this was a cold front, and a mass of cold air was moving toward us.

Obviously that cold air was moving along under the warmer and lighter air in its path. On the map, too, I could see that the cold front ran through the middle of two ovals in South Carolina and Georgia. Those ovals indicated a swirl of air which goes with a storm.

The storm, I could see, was not far from the Atlantic coast. Even in February, one could count on the air temperatures over the ocean off South Carolina being 15 to 20 degrees warmer than air temperatures over land. If warm, moist ocean air began flowing into the air mass, the storm would grow rapidly.

The storm could be expected to move, in normal fashion, northward up the coast. It could possibly move out to sea—or it could possibly hand New York City and the northeast a heavy snowfall.

The map, of course, showed what was going on now. I had to predict what could be expected tomorrow. I forecast heavy snow. The weather service rested with a forecast of "snow beginning in the early morning hours, changing to rain. Rain ending in the early afternoon."

We got 15 inches of snow.

As I told an interviewer at the time, "To some extent, I was lucky. A lot of pieces of information go into making a forecast, they're always changing, and it's easy to slip up. With that storm, for example, there were many questions: How fast was it going to move? Was it going to get stronger or weaker? What was the temperature going to be when it got here? One degree below freezing and there could be a bad snowstorm. One degree above, and we'd get nothing but rain. In this case, the weather service thought the storm would move faster than I did. They also thought the storm would be weaker and the temperature warmer."

At other times I've been wrong and the weather service right.

There was the time that I predicted a chance of showers for the next day. The National Weather Service called for heavy rain. I thought the coastal storm would slide by to the southeast. It didn't! Well, we had record-breaking rain and serious flooding. It took me a while to live that one down.

TV weather forecasting is a game. It's a job but it's fun. And it's fun and a game because there has to be an element of gambling in it.

I probably don't have to tell you—you know from experience—that some guesswork has to go into prediction and the guessing isn't always good.

If you're a forecaster, you can pretty well count on going to sleep hopefully and waking many mornings wondering, slowly opening one eye to see, is it, as you predicted, snowing? No, the sun is coming up and the sky is clear. And you wonder what happened.

So you call the National Weather Service and ask: "Where in the devil is that low we were supposed to be getting?"

And the answer may be: "It's still deepening, hasn't moved much, but it should begin to move up this way pretty soon now."

You blew it. In this case, by only a couple of hours and by the time you go on the air later in the day, the snow will be there and you'll be out of trouble.

You can count, too, on there being other times when you've predicted a beautiful, sunny weekend and wake on Saturday to find overcast, drizzle, wind gone into the northeast. And you know

49

the front that should have come through did not, is still to the south, and the area is socked in for the weekend. There go the lovely, wonderful couple of days you've been telling all those nice people about and for which they've raced out to Fire Island or wherever.

At which point you tell yourself: "I'd better not go shopping today. I'd better stay home." Because, invariably, if you walk out on a weekend like that, and if you go into a store looking for some tools or anything else, fifty people will come up and say, "Nice weekend, eh?" And you're a dead duck, at least for then.

You can be right nineteen times and nobody says boo. But when you blow one, count on the "What happened to the nice weekend?" slap.

Slap? I remember one occasion when I really blew a forecast and it was serious. We had a mammoth snowfall and I had predicted just a chance of flurries. When I went on the air, I said: "Let me first apologize about the fact that you didn't get flurries but 12 inches of snow." At that point, about twenty snowballs were thrown at me. One of them hit me neatly on the side of the head and for a moment I was stunned.

It was the stage manager who had gone out, scooped up some snow, and made the tight snowballs. He almost got fired. I had to go in and plead for him with a lot of assurances that I really wasn't too upset.

The viewing audience, I'm sure, loved it. Probably most of them would have liked the opportunity to throw a few at me.

It's almost impossible to forecast with pinpoint accuracy in many situations.

Scattered showers? That's what they'll be—but you can't predict the scatter, whether, for example, one might hit Yankee Stadium or not. It's like spilling a few drops of water off the top of a building. You know the water is going to hit the ground and, if the wind is blowing out of the north, it will land south of you—but how far south?

I don't know how many times I've gone on the air during a winter storm and said: "So for tonight rain will continue in the

50

metropolitan area, heavy at times; the northern suburbs will have some freezing rain, sleet and even some snow." And not long afterward the phones will go crazy with people calling in to say: "Rain? You're nuts. We have 12 inches of snow." And they do—just 50 miles away from New York. But in midtown Manhattan it's raining and continues that way. And for city dwellers the next morning, when folks drive in from the suburbs, the amazing sight is the 12 inches of snow on top of auto roofs on the city's dry streets.

I've had my share of complaints and even a snowballing or two. But—so far—I haven't had the unhappy experience of one TV weatherman in Louisiana who got conked in the middle of a broadcast. A viewer, furious over a bad prediction the day before, had driven to the station, somehow managed to get himself up into the studio and over he marched to the unsuspecting weatherman and knocked him cold with one punch.

Many strange things happen to weathermen. On New Year's eve 1977, Winston the Weatherdog lost his position as my assistant on NBC-TV in New York. NBC vice-president Joe Bartelme ordered that Winston was to be barred from the early and late news broadcasts. That firing was on the front page of the New York *Post* the next day and on the AP wire service. The battle over Winston raged for several weeks, during which we received thousands of letters of protest about Winston's dismissal.

It all began when my son, Storm, brought Winston home as a gift for my birthday. Winston was not housebroken, and because I was living in a small apartment in New York City, I decided to bring Winston to the office with me.

My office was covered with newspapers. Tom Snyder dropped by while I was out and fell in love with Winston. He had a brainstorm, and that evening when he introduced the weather segment, he said, "Hey, understand you have a bulldog pup now, Frank." I had no idea he knew Winston was in the building. "You know, Frank," Tom said, "I have always loved bulldogs. Bring him down when you finish the weather."

I completed the weather report, and Tom went on with the

news while I ambled upstairs and brought Winston to the set, waiting for the program to end so Tom could see him.

However, immediately after introducing Tom O'Reilly, the sportscaster, Tom Snyder began talking about Winston and asked me to bring the pup on camera.

Everyone oo'd and ah'd at the beautiful soulful expression on Winston's face, and Tom said, "Now tell us, Frank, why your office smells like a kennel and why Mrs. Field won't let you keep Winston at home."

As I tried to explain, the switchboard lit up like a Christmas tree, with viewers expressing the gamut of emotions. Tom's office was flooded with calls. Thus began the career of Winston the Weatherdog, a full-blown TV personality who chewed on my shoe while I was presenting the weather, or barked when I said rain.

Winston received presents daily: Valentine cards, a 50-pound sack of dog food, a tooled-leather choker, something called a dog-a-teria and even offers to appear on commercials. But I turned those down since Winston was a newsdog and was not permitted to do commercials.

Winston's downfall was the result of a special invitation to attend a New Year's Eve party.

Carol Jenkins, one of our reporters, visited with the late Guy Lombardo, who was celebrating his fortieth anniversary of ushering in the New York at the Waldorf Astoria. Guy invited Winston to join him in his closing comments. It got the laugh in the studio, but I, on behalf of Winston, accepted the invitation to the New Year's party and promised the audience that Winston would appear that evening on the eleven o'clock news to model his sweatshirt tuxedo, a present from a viewer.

Among the hundreds of phone calls we received after that announcement was one from our VP of news, who told our anchorman, in essence, that Winston would appear only over his dead body.

At 11:00 that evening I apologized to the audience and explained that Winston had been barred—but I held up his little tuxedo and said he would be wearing it to the New Year's party the next night.

The switchboard went mad. Viewers were livid. One call was from the AP. The caller thought it was a put-on, but, when he discovered it was for real, he traced the story back through various sources at NBC. Since it was New Year's Day and a slow news day, it was a perfect feature story, and out it went with pictures on the Associated Press wires. Papers all over the country picked it up and Winston's plight was even lamented in Hong Kong and Rome.

Winston never came back. His feelings were hurt. As for me, it could not have happened to anyone but the weatherman on a news program.

Viewers respond frequently by mail or phone. You get some weird or funny calls. I got a call one night just before going on the air and I had difficulty keeping a straight face when I did go on. The call was from a drunk. "Scushe me, doc," he said, "but wha's the weather from'ere to Chicago?" And I said: "You mean the flying weather?" And he said: "Oh, thank you very mush."

I had one woman who almost drove me crazy with her calls for almost a year after we first introduced radar on the broadcasts. She kept insisting that I was watching her with my radar and she couldn't take a shower because she knew I was watching her, and she was going to call the FBI and the CIA and was writing to the president of NBC News.

One very funny incident—that is, funny in some ways but not in others—occurred one night because of a stagehand's mistake. In those days, we had a rear projection machine to throw the weather map up behind us. This night, the man behind the screen doing the projecting put the map up backwards—New York was on the west coast and Los Angeles on the east.

Duly, I approached the screen, said, "Let's look at the map," turned to it and was bewildered. I wondered if something was wrong with me. But I recovered quickly and said, on the spur of the moment, "Well, you'll notice that our map looks a little different tonight but that's all right—if you turn your set the other way, it will work fine," and I quickly went on.

Well, NBC lawyers got a letter from a viewer who had gotten up, turned his set, knocked a lamp over and broken it, and wanted us to pay him for the lamp.

I'm sure there isn't a TV weather person anywhere who hasn't had his share of bloopers and other funny or weird experiences.

Miss Essie was the first weather girl in Kansas City, Mo., on station KCMO. I know about Miss Essie because her son, Chuck Scarborough, is now the anchorman for NBC's evening news broadcast in New York and he told me the story.

Miss Essie had a unique way of delivering tomorrow's weather. She had a large laundry basket out of which she selected clothes to hang on the studio wash line. The shorts would say "Fair tomorrow"—the shirt would specify the temperature. Miss Essie was a popular and beautiful weather girl.

To capitalize on their sponsorship of the weekend weather segments, the Union Chevrolet dealer ran a large promotion campaign. Bring your car in early Monday and Miss Essie, the beautiful weather girl, will drive you to work while your car is being serviced.

The problem was—that Miss Essie had never driven a car in her life! When the station and the sponsor learned that, they rushed Essie into a crash driving course for a week, pulled strings and got her a license just in time for that momentous Monday morning.

The first customer arrived and Miss Essie and the photographers greeted him at the auto dealer's. After publicity pictures were taken, the customer slipped into the large white station wagon. On the roof was emblazoned the name of the popular weather girl—on a giant four-foot thermometer.

Off went Miss Essie down the road—jumped a traffic light at the next corner and plowed into the midsection of a florist's truck, strewing carnations all over the neighborhood! Miss Essie's weather career lasted all of five weeks!

Then there's the acrobatic weatherman. He's Barry Lillis of WGR-TV, Buffalo, New York. He literally leaps and spins through his weather report. He remembers the night he was delivering the late weather, hurling a high or a low at his map, slamming the temperature board, bouncing about on camera, when he suddenly became aware of someone apparently doing the Mexican hat dance offstage. The man was hopping up and down on a fouled-up microphone cable, which had snagged on an extremely hot stage light. It was Barry's mike cord and it was on fire. The rest of the weather was delivered as Marcel Marceau might, in complete pantomime, with acrid smoke drifting across the cameras.

And look at what happened to Willard Scott. After attending a function between broadcasts, Willard arrived back at the studio in time to go on the air. Anchorman Jim Hartz tossed Willard the cue, and off he went, never at a loss for words.

But what Willard didn't know was that someone, perhaps a stagehand or a young visitor on the set, had been tinkering with his map. It was not until he was well into the report that Willard realized that the temperature at Miami, which read 14 degrees with 11 inches of snow, couldn't be correct. Nor would Caribou, Maine, have a reading of 89 degrees during February. Nor would the winds be circulating the wrong way around lows and highs. For the first time, the unflappable Willard Scott was flapped. Even today Willard won't talk about that particular weather report.

Then there's the one about the weatherman who got a new radar screen on his set. For almost three weeks he pointed out the heavy showers over the Blue Ridge Mountains about 50 miles west of his city. But the folks in that area never saw a drop of rain. Someone at the National Weather Service mercifully put an end to those daily showers by pointing out to the errant weatherman that those blips on the screen were merely reflections of the mountains. The blips are still there each night but there are no longer showers.

One weekend I watched my son, Storm, at the competing station WABC-TV deliver his weather report. At the time Storm was employing a metallic weather map. The symbols were applied by hand, each shower or front or pressure system adhering to the metal weather map with magnets. But this evening I watched Storm's confidence slowly give way to horror as he realized that the weather symbols were not sticking. The shower he placed over the Great Lakes was slowly sliding down into the Gulf. The cold front on the East Coast was dangling into the Atlantic and from there it went crashing to the floor. You see, the map had been repainted and lacquered . . . and only chewing gum would hold. I had to tune out. It was bad for my heart.

5

The Weather Machine

THE "HIGH" is building over us. Our barometric pressure is falling, and the "low" heading this way will bring rain. You've heard these expressions time and again in weather reports on TV and on the radio.

The jargon of weather also includes terms such as "jet stream," "relative humidity," "windchill factor," "northeasterly flow," "prevailing westerlies." And if these and many other terms descriptive of what makes up our weather machine seem more mystifying than (or just as baffling as) the jargon of sports reporting, which employs "hat trick," "sweeping curve ball," "off-tackle slant," "dunk shot" and so on, they need not be.

With a little pertinent delving here into the elements that make up the weather machine, chances are you'll soon find yourself understanding not only more of the technical terminology but why weather does what it does. And you may even find yourself arguing, enjoyably, with your local TV forecaster over how he goes about making his forecasts, a process we'll look into in the next chapter.

Not likely? More than once, I can assure you, I've had a sophisticated viewer call me up and say, as one did for example, "Hey, for gosh sakes, didn't you see on your satellite photo that you had this mass south of New York—and anyone could have told you that the mass was growing because it was smaller yesterday, but you chose to ignore it." And I then have to explain that, yes, I did know it was there but thought that the northeasterly flow of air would keep us dry and we would not have any problem—and yes, he was right, and I was wrong.

We live at the bottom of a colorless sea of air, a huge mass of some 5 billion million tons that surrounds the earth, covering land and sea, rotating with the earth as the earth orbits the sun, but having a "mind" of its own, a circulation of its own relative to the earth's surface.

Our air, or atmosphere, is a mix of gases. At sea level, air contains about 21 percent oxygen, 78 percent nitrogen, and about 1 percent argon, plus very small and varying amounts of other gases such as carbon dioxide, helium, krypton and neon.

There's something else in the air and it's of major meteorological importance: water vapor. Without it, we'd have little if any of the phenomena known as weather. In fact, without it, we'd have no cloud cover for protection against the sun, no rain for watering plant life—no life at all, at least as we know it.

Virtually all water vapor occurs in the air below the 25,000-foot level. And in that lower layer of the atmosphere, too, are solid particles—dust, smoke, salt from sea spray—which range in size from submicroscopic to bits large enough to be visible to the eye.

In meteorology, the atmosphere is considered to have two major regions, lower and upper. And the lower layer, which is the stormy region, is called the troposphere, from the Greek *tropos* meaning to "turn or mix." Its thickness varies, averaging about 54,000 feet over the equator, about 28,000 over the poles, and about 37,000 feet over the temperate zones.

It's in the troposphere, the lower layer—with its vapor and particles and vigorous mixing of air—that most cloud activity occurs and all air masses, fronts and storms develop. It's the scene of weather in all its manifestations, as we know them.

Imagine sitting at the beach on a sunny day. You face the sun and you quickly feel the solar rays burning the skin on your face and chest. Your back and the top of your head are relatively cool. Those parts of your body exposed to the sun begin to heat up. To get a nice even tan you slowly change position.

Well, the earth, as it slowly rotates, also captures solar energy in such a way as to get an "even tan!" It is at the equator, the point of greatest exposure, that the solar energy impacts and the land and ocean surfaces receive their maximum heating.

From experience you know that warm air rises. That is what sets our weather machine in motion.

When heated, air molecules speed up their movement and tend to draw away from each other. As a result, the heated air expands, its density decreases, it weighs less than surrounding cooler air and it rises.

And, overall, there is a tendency for hot tropical air to move upward and toward the poles and, after being cooled there, to move down and back to the equator to be heated again. Thus the poles are less cold than they otherwise would be and the tropical areas less hot.

But there's more to the story.

Winds, of course, are important determinants of weather and our major, large-scale winds—global air movements—develop because of temperature differences at the equator and the poles, and also because the earth rotates.

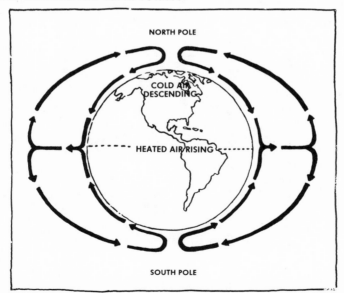

Theoretical air movement if the earth did not rotate.

59

To see how that happens, we start with the area extending about 10 degrees on each side of the equator, an area known as the doldrums, or the equatorial zones. Here, the great columns of heated air rise. As they do so and reach toward the top of the troposphere, they cool and become heavier again and would tend to sink. But they can't because still more heated air, getting its heat from the sun-baked lands and waters of the equatorial region, keeps moving up.

The higher air has to move somewhere and it does—horizontally—some toward the North Pole and some toward the South Pole.

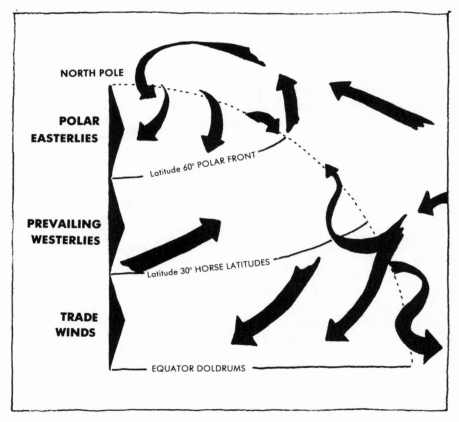

The effect of the Coriolis force.

As the flows—northbound and southbound—reach latitudes about 25 degrees north and south of the equator, some of the air sinks toward the earth, producing areas of high pressure. As it sinks, the air is warmed by compression. These are areas of calm hot air, clear skies and little rain known as the "horse latitudes"— perhaps because in early days windjammers transporting horses lost considerable numbers of their cargo to the heat.

Near earth in the horse latitudes, the descending air branches and one stream moves poleward while the other flows back toward the equatorial zones. That would be our simple air circulation!

But the rotation of the earth has a significant influence. It deflects the streams of air.

That deflecting effect was discovered by and named after a nineteenth-century French civil engineer, G. G. Coriolis. Although many complex explanations can be offered, it boils down to this: Because the earth spins toward the east, all moving objects in the Northern Hemisphere tend to turn somewhat to the right of a straight path while those in the Southern Hemisphere turn slightly left.

Let a pilot set out to fly, say, from Seattle to New York, he has to compensate for the Coriolis effect or his plane will end up in South America.

Because of the Coriolis effect, the stream of air flowing from the horse latitudes toward the equatorial zone veers to the right and forms the mild, gentle, steady trade winds that blow as northeast winds in the Northern Hemisphere (and which carried Columbus to America) and as southeast winds in the Southern Hemisphere.

Now let's pause for a moment. In meteorological language winds are described as to the direction they blow *from*. A northeast wind is blowing *from* the northeast. A southeast wind describes air that is flowing from the southeast toward the northwest. The air moving poleward from the horse latitudes also veers (because of the earth's rotation) to the right in the Northern Hemisphere and left in the Southern Hemisphere. The result: the prevailing westerly (air moving from west to east) winds found in both hemispheres in the latitudes between 35 and 60 degrees.

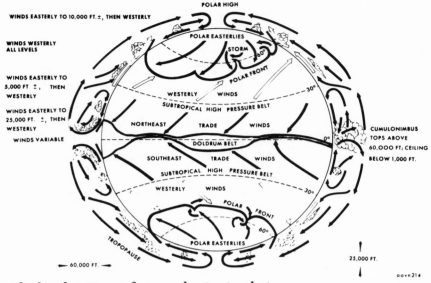

WINDS EASTERLY TO 10,000 FT.±, THEN WESTERLY

WINDS WESTERLY ALL LEVELS

WINDS EASTERLY TO 5,000 FT ±, THEN WESTERLY

WINDS EASTERLY TO 25,000 FT. ±, THEN WESTERLY

WINDS VARIABLE

POLAR HIGH

POLAR EASTERLIES

STORM

POLAR FRONT

WESTERLY WINDS

SUBTROPICAL HIGH PRESSURE BELT

NORTHEAST TRADE WINDS

DOLDRUM BELT

SOUTHEAST TRADE WINDS

SUBTROPICAL HIGH PRESSURE BELT

WESTERLY WINDS

POLAR FRONT

POLAR EASTERLIES

TROPOPAUSE

CUMULONIMBUS TOPS ABOVE 60,000 FT; CEILING BELOW 1,000 FT.

60,000 FT.

25,000 FT.

aavn214

Idealized pattern of atmospheric circulation.

You'll recall that only some of the air from the equatorial region sank at the horse latitudes, resulting in the trades and the westerlies. Some continued on at high altitudes to the poles, reaching there, cold and dense, piling up and producing a dome of heavy, frigid air at each pole. And this cold, dense air, pushed by arriving new quantities, starts to circulate back to the equator but, because of the earth's rotation, is deflected somewhat to the right. It becomes the polar easterlies, blowing in the Northern Hemisphere from a northeasterly direction and in the Southern Hemisphere from the southeast.

The middle latitudes where we live are where the very different air masses—the cold of the polar easterlies and the warm of the prevailing westerlies—meet, producing atmospheric instability and playing a major role in creating the changeable weather of our temperate zone.

AIR MASSES

It's winter. Each evening your favorite weather personality delivers the same story. A "mass of arctic air" will continue to dominate our region and temperatures will remain below normal. Where did that cold air mass come from? Or it's midsummer and it's the heat that's making the headlines. And this time the blame is placed on a "tropical air mass." Where did it come from and where is it going?

Just imagine a huge mass of air like a giant drop of molasses covering all of northern Canada around the arctic. Day after day this mountain of air sits over ice and snow being chilled by the frozen land surfaces. Any warmth in this air mass is lost to space, radiated upward through clear cold skies. Soon, the giant air mass is uniformly frigid and begins to slide southward heading towards the United States.

Wherever you are, the weather you experience depends either on the character of the air mass prevailing at the time or on the interaction of two or more different air masses.

An air mass is a huge body of air thousands of square miles in size with much the same characteristics—dry and cold or moist and warm, for example. Air masses are formed in certain areas of our hemisphere known as source regions, where air tends to stagnate for days and weeks at a time, giving the source regions a chance to impart their native characteristics of temperature and moisture.

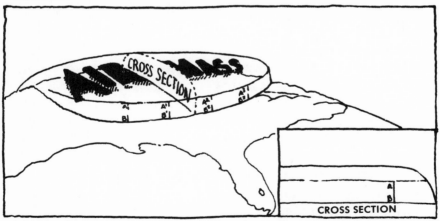

Horizontal uniformity of an air mass. (Properties of air at A^1, A^2, etc., are about the same as those at A; properties at B^1, B^2, etc., are about the same as those at B.)

Four different types of air masses are important to us. You'll find them identified thus on weather maps: cP for polar continental; mP for polar maritime; cT for tropical continental; and mT for tropical maritime.

Polar continental, which originates in the arctic, particularly in north-central Canada, Russia and Siberia, is cold and dry. In summer, when it moves south into the central and eastern United States, it brings relief from heat, with cooler temperatures, low humidity, northwesterly winds, and clear skies. In winter, it can sweep in to produce a cold wave, lowering temperatures tens of degrees within a few hours.

Polar maritime air masses develop over the northern Pacific and Bering Sea and over the colder regions of the Atlantic Ocean. Those from the Pacific, which are somewhat warmed by the Japanese Current, are mild and moist and bring a moderate climate along with some precipitation and fog to Pacific Coast areas such as San Francisco, Seattle and Portland. Those from the Atlantic, bringing low to moderate temperatures, high humidity and much cloudiness, often move into the northeastern United States, especially in fall and winter.

Tropical continental air masses, which originate over Mexico and the southwestern United States in summer, are warm and dry—and occasionally, when excessive, can produce drought.

Tropical maritimes, warm and laden with moisture, are produced by the sun's heating of such tropical waters as the Gulf of Mexico, Caribbean Sea, the Sargasso Sea area of the Atlantic, and the mid-Pacific. In summer, they tend to prevail over much of the eastern part of the United States, bringing high temperature and high humidity. In winter, when they are often shoved up over polar continental or polar maritime air, they cause cloudiness and rain. Tropical maritimes from the Pacific have fairly high humidity and moderate temperatures. They occur along the Pacific coast and upon reaching the mountains produce much of the frequent Pacific Northwest rains.

Television weathermen may, according to the area for which each is reporting, allude to the same air mass in different terms:

frigid arctic air mass, Canadian air mass, tropical Gulf air, Caribbean air mass.

An air mass changes as it leaves its original source and travels. Moving across mountains, oceans and rivers, it begins to acquire new character. For example an arctic air mass is slowly warmed as it moves southward. The subzero weather it brings to the Plains slowly undergoes a change. By the time the huge mass rolls into the Southeast Atlantic states it is no longer the same air mass that gripped the Canadian provinces. The same air mass which brought 5-below-zero weather to Fargo, North Dakota, a week ago is now chilling the Florida Panhandle with temperatures in the 40s. And a few days hence the cold air will be out over the Atlantic producing a dry cool nippy weekend in Bermuda.

FRONTS

Now let's deal with one of the most used and abused of weather terms—the front! It's really quite simple to understand.

A front is just that—the front, or leading edge, or boundary line, of whatever air mass is moving your way. So a cold front is the forward edge, or front, of a huge mass of cold air. A warm front is the boundary, or line, that marks the leading edge of a mass of warm air.

It's obvious that when different air masses meet, a clash will ensue. And it is in the middle latitudes where cold polar air is moving southward and warm tropical air is moving northward that the two continually interact with each other. The cold air is the aggressor. It's leading edge is moving, and in the winter the zone of interaction is called the polar front. It is not stationary. At some places, the polar front advances south as a strong flow of cold air pushes southward and replaces the tropical air. At other places, the front can be moving northward ahead of advancing tropical air.

Sometimes, in winter when cold air masses are dominant, the polar front can move well into the tropics. In summer, when warm air masses dominate, the front may retreat as far as 60 degrees north.

65

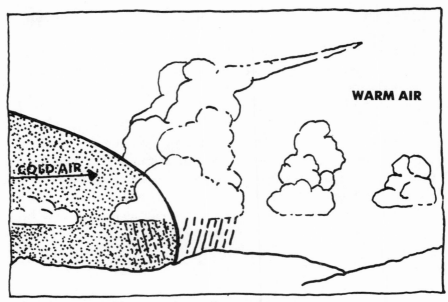

A cold front in vertical cross section.

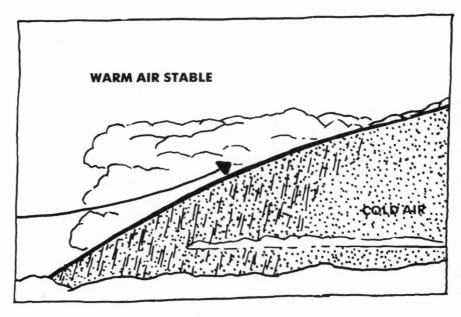

A warm front, with stable warm air.

The polar front is the main zone of clash. But fronts may form between any air masses that are dissimilar.

At any one time, there can be dozens of air masses, old and new, some almost sitting still, others moving hundreds of miles a day, moving most rapidly when they are in the westerly airflow in the middle latitudes.

Fronts can be either warm or cold. If the cold air mass is retreating and the temperature goes up as the front moves over an area, the front is called a warm front. If warm air is retreating and cold air advancing so the temperature goes down over an area, the front is called cold.

It might be better, in fact, if fronts were labeled warmer and colder. Temperatures are relative. On a hot summer day, for example, a cold front may have a temperature of 70 degrees behind it, hardly cold but nonetheless cooler than the hotter air ahead of it.

Along a warm front, the advancing warm air is lighter than the cold air ahead and the warm air rides up over the cold even as it exerts some push against the cold air at the earth's surface. There is thus a long, gradual slope, with the warm air moving up the slope being farther advanced than the warm air at the surface. This is why a warm front can announce its coming well in advance, as much as one to three days before you actually feel the warmth.

As the warm air rides up over the cold in the long ascent, it cools. It then can hold less moisture than it did when warmer, and the excess moisture condenses into clouds, which may stretch as much as 500 to 1,000 miles ahead of the warm front's surface position. At first, thin cirrus clouds form high up, then gradually descend and thicken, producing rain in a belt that can often be as much as 300 miles wide ahead of the front.

As the front passes, temperature begins to rise, wind shifts, clouds thin and break up, and rain stops.

With a cold front, the cold air, being denser, hugs the ground and shoves up the warm air ahead. The warm air rises in a much steeper slope than with a warm front, and, because of that, the changes come much more quickly, with much less warning, and are more violent.

If the cold front is moving rapidly and the air ahead is moist, there are likely to be gusty winds and heavy but brief rain. Clearing is rapid; the wind shifts, blowing from the cold region, temperature falls, pressure rises, and humidity lessens as the front passes. But if the front is moving slowly and displacing stable warm air, clearing is likely to be quite slow.

Sometimes fronts lose some of their momentum. For a period, they appear not to move. They are then called stationary. The weather associated with a stationary front may be like that associated with either a warm or a cold front or sometimes marked by just a belt of clouds.

There is also what is known as an occluded front. When a cold front overtakes a warm front, the warmer air is lifted, and cold air behind the cold front can meet with cold air ahead of the warm front. There is likely to be some difference in temperature between the two masses of cold air. If the air behind the cold front is colder, it will lift the relatively warm air ahead of the warm front. If the air ahead happens to be colder, it will lift the air behind the cold front. An occluded front usually means stagnant, poor weather.

To just slightly complicate matters concerning the weather produced by fronts—there are frontal systems which may be accompanied by nothing more than a change in temperature and humidity, without clouds and precipitation. For example, a dry air mass may replace a comparatively dry mass already over your area. The passage of the cold or warm front may be noted by changes in barometer, a shift in wind, a change in temperature and a variation in humidity but without inclement weather.

When you watch your television weather report, you may see fronts depicted on the map in their usual form or in an artistic version.

HIGHS AND LOWS

Air masses, whether they are cold or warm, may build up into huge mountains of air that are quite heavy. With that mountain of

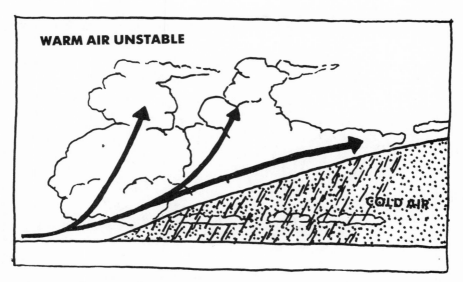

A warm front, with unstable warm air.

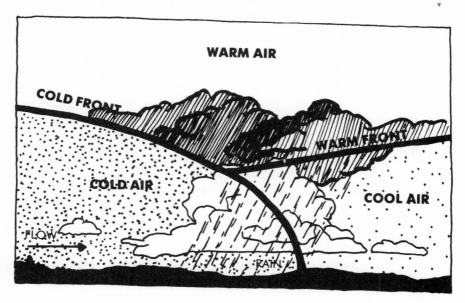

A cold-front occlusion in vertical cross section.

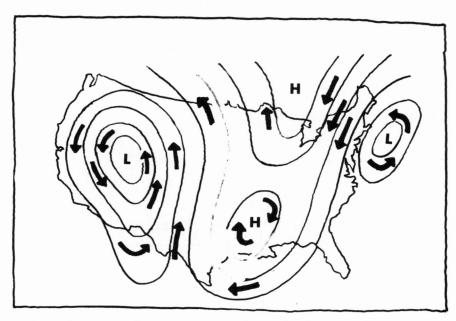

Flow of air around pressure areas above the frictional layer.

air on top of us, we should feel the pressure—but the human body for the most part accepts and scarcely notes it. There are devices that can register such variations in pressure, and as you probably know a high barometric pressure often is associated with good weather, while a falling barometer heralds an approaching storm.

What the barometer is recording, of course, are areas of high and low pressure in the atmosphere as they move in your direction.

In a high region of the atmosphere, air settles downward, producing winds that blow clockwise and slightly outward from the center. Because of the settling, the air is heated by compression and this has a cloud-dissipating effect. Usually, with a high, the weather is fair; so when your weather forecaster offers his or her nightly statistics and points out that the barometric pressure is rising you have a clue as to what to expect. High pressures are sometimes referred to as anticyclones.

On the other hand, a falling barometer means that the mass of air over you is bringing lower pressure. If that pressure continues to lower over a period of time then you know that a low pressure system is approaching.

In an area of low pressure, also called a depression, air moves inward toward the center, creating a wind that runs counterclockwise. Generally, lows bring unsettled conditions and storms.

In the United States, there are alternating processions of highs and lows, with the highs commonly moving east and southward and the lows east and northward.

Many lows enter the country in the far northwest. Some move east along the northern border across the Great Lakes and out over the North Atlantic. Some move toward the south in the center of the country and then curve back up north and east. Still other lows develop in the area of the southern Plains or move into the southern states from the Gulf of Mexico and move northeast and exit off the north Atlantic coast. A high typically moves into the United States from Canada east of the Rocky Mountains and then moves southeastward.

Areas of low pressure with their counterclockwise circulation are known as cyclones (not to be confused with the severe storms to be mentioned later). Low pressure systems may cover many

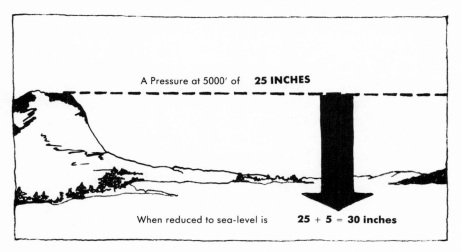

A Pressure at 5000′ of **25 INCHES**

When reduced to sea-level is   **25 + 5 = 30 inches**

Reduction of station pressure to sea level.

71

thousands of square miles, the air moving counterclockwise and at the same time flowing inward toward the center and rising like a huge whirlpool in the atmosphere. Lows can either drift slowly across the country or deepen and move faster. The speeds at which both low and high pressure systems move vary greatly from just a few miles a day to 500 miles a day.

THE SEMIPERMANENT HIGH- AND LOW-PRESSURE
CENTERS

Look at a typical weather map of the Northern Hemisphere and you see two large high-pressure ridges, one usually over the middle latitudes of the Pacific and the other of the Atlantic, each often reaching out with an arm near to the closest U.S. coastline.

Known as the Pacific and the Bermuda or Azores highs, they are two of several main generators of air movement which, because of their persistence, are called semipermanent cells. All such cells exert major influence on weather.

The Pacific high, which covers millions of square miles, has a somewhat changeable position approximately north and northeast of Hawaii. As cyclonic lows move across the Pacific, they are steered around the north edge of the Pacific high. The Pacific high's northeastern arm, depending upon where it is located at various times and its strength, has much to do with determining the latitude at which storm systems crossing from Japan and Siberia penetrate into the North American continent.

In the Atlantic, a similar but somewhat smaller high has its center in the area of the horse latitudes and stretches across the ocean from near the Middle Atlantic states coast toward the Spanish coast. It is known generally as the Azores high while its western extension, of great interest to us in the United States, is called the Bermuda high.

The Bermuda high drives winds to our east coast. In summer, when the high tends to expand, it gives the east coast its prevailing summertime winds from a southerly quadrant, and brings moist warm air from the Gulf of Mexico as well as the Atlantic to much of the eastern part of the country. It often blocks the flow of cool dry air from the west and north. The Bermuda high also influences the path of hurricanes from the Atlantic.

Two major low-pressure areas that persist most of the year also have much to do with our weather. One, the Aleutian low, shifts about somewhat in the North Pacific Ocean not far from Siberia and Alaska. Here arctic and tropical airstreams meet to produce an area of great storminess. And not infrequently, breaking away from the Aleutian low, smaller but still potent lows move eastward across the continent, sometimes in groups of three, four and more, affecting weather for several weeks.

The second low, called the Icelandic low, between Greenland and Scandinavian coastlines, spins cold air from the northern Atlantic to Canada and to Scandinavia, Britain and northwest Europe. When it sometimes moves near Newfoundland, the Iceland low can influence air circulation over New England and eastern Canada, intensifying offshore storms and making for long periods of northerly gales.

JET STREAM

"The jet stream will steer the storm center north of us." You hear quite a bit about jet streams when you watch TV weather forecasters around the country. In fact some weather presentations often feature the position of the jet stream on the video weather map. It resembles a river of air in arrow form on the colorful national weather map. It's popular today but no one knew the jet stream existed fifty years ago.

During World War II, American B-29s bombing Japan from 30,000 feet encountered unexpectedly strong westerly winds. Their aim was spoiled.

Until planes flew that high, the existence of such winds with speeds of up to more than 200 knots was unknown. Later, these winds got the name jet stream.

There are several jet streams, rather than just one. A jet stream can be continuous around the hemisphere but more often it is

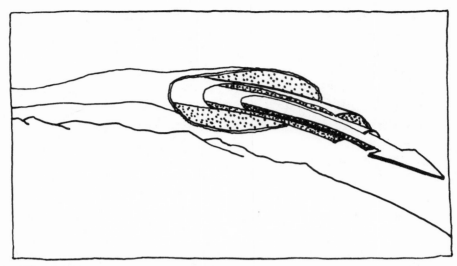
The jet stream.

broken into disconnected segments that shift north and south as well as vertically. Jets can be thousands of miles long, hundreds wide and several thousand feet deep. Commercial airlines often take advantage of the jet stream to save expensive fuel and gain time by allowing these strong winds to sweep them eastward. Conversely, pilots will often fly off course to avoid encountering the jet stream headwinds.

Among the jet streams, the one of most importance to the United States is known as the polar-front jet stream. Its strength is greater in winter than summer. Its mean position shifts south in winter and north in summer just as does the polar front. In winter, it can shift as far south as 20 degrees north latitude.

Sometimes the position of the jet stream can be judged from clouds that form below it. They are usually drawn out in longitudinal streaks in line with the direction of the high winds.

In winter, then, the polar-front jet stream is almost always over the United States. In summer, it moves back up over Canada— but then in the southern states we get a bit of another jet stream, the subtropical, which moves somewhat north.

High up as they are, jet streams can markedly influence wind and weather down below. They bring together air masses coming from different regions and the results on the ground can be much like those produced by air-mass antics triggered lower down.

The unusually snow-free winter of 1979–1980 in the northeastern United States is attributed to the shift southward of the controlling jet stream. Instead of storms tracking in their usual direction out of the center of the nation and from the Gulf of Mexico toward the New England states, the systems were shunted through the Middle Atlantic and southeastern states. The result was more snow in the Virginias than in the normally snowy northeast.

LOCAL WINDS

In large metropolitan areas the media weather forecaster faces a problem brought about by local winds. When the forecaster says, "Readings will range from a high near seventy at the shore to

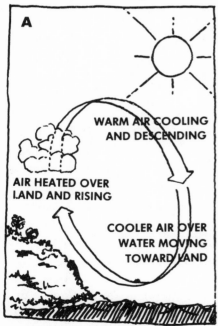
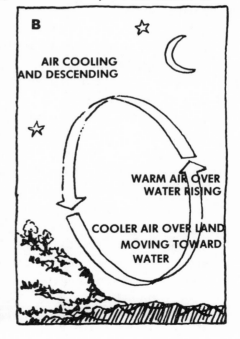

Land and sea breezes.

eighty in the northern suburbs to near a record high of ninety-three here in the city," it may sound like a hedge. Why the disparity in temperatures?

Important as the general wind and pressure systems are, weather in an area can be influenced by local pressure and wind systems created by mountains, valleys and water masses.

Land and sea breezes. Land warms and radiates heat more rapidly than does water. So, in coastal areas, the land is warmer than the sea during the day but colder at night than the slower-to-lose-heat water. The temperature variation produces a corresponding variation in pressure, and by day, pressure over the warm land is lower than over the colder water.

During the day, then, warm air over the land will rise to higher altitude and move horizontally out to sea. To replace the rising warm air at the surface, colder air over the water moves onto land. The result is a sea breeze that begins to spring up in the morning, increasing until the warmest time of day, afternoon, then dying down at sunset.

At night, the circulation is reversed. A land breeze springs up as the sea stays comparatively warm, the land cools, and the cooler, heavier land air moves out to sea.

Sea breezes usually are stronger than land breezes but even so, seldom move far inland. Both land and sea breezes are shallow in depth.

For those who live in coastal areas of the United States, it's a never-ending source of amazement to hear of record-breaking heat inland while along the coast it's been sweater weather. Or to learn that the rain that fell along the coast came down as snow just a few miles inland where temperatures were just a few degrees colder.

Mountain and valley winds. In mountainous areas, similar uneven heating breeds winds. On a warm day, air touching a mountain slope is usually warmer than air farther from the slope. During the day, the slope-touching air becomes lighter than the surrounding air and rises up the mountain. This movement of air is often called a valley wind since it seems to be moving up out of the valley.

At night, as the mountain slope radiates away its heat, air in contact with it becomes colder and denser than the surrounding air and sinks. The result is the mountain breeze, so called because it seems to move down from the mountains.

Foehn/Chinook winds. When damp air arrives at a mountain range, it must rise. As it does so, clouds form and rain develops on the windward slope as the rising and cooling air gives up its moisture. For every 1,000 feet of rise, the temperature of the damp air falls 3 degrees Fahrenheit.

Forced air flow over mountains results in the "chinook."

When the air, now free of much of its moisture, arrives on the leeward slope and descends, its temperature rises—and it does so at a rate of 5½ degrees Fahrenheit for every 1,000 feet of descent. Thus, if the air rose, say, 10,000 feet and descended the same distance, it arrives at the bottom of the leeward slope 25 degrees warmer than when it was on the windward side at the same level.

The name given to this warm downslope wind is foehn in the Alps and chinook along the eastern slopes of the Rocky Mountains and its associated ranges.

6

Putting Together a Forecast: The Behind-the-Scenes Story

WHAT GOES into a weather forecast? What happens behind the scenes before someone like me steps in front of a TV camera, points at maps and photos and announces with more or less conviction that "Tomorrow. . . "?

Forecasting, once a folk art (and sometimes a fairly accurate one), today is a burgeoning science called meteorology that uses an ever-growing array of technological tools—radar, satellites, balloons, computers—in a vast information-gathering network.

The basic information upon which tonight's weather forecast is based is derived from one source—the government's weather service. Private meteorologists, TV, radio and newspaper weather reporters all depend on that same network of information.

THE NETWORK

Each day, into the U.S. National Weather Service Meteorological Center (NMC) in Suitland, Maryland, comes a huge amount of data about the weather.

The information flows in from thousands of observing stations, manned and unmanned, all over the Northern Hemisphere. Some 500 of those stations are in the United States and North America. Additional information comes via more than 3,000 ship reports on weather during an average day, as well as some 1,000 from commercial aircraft.

Four times a day, at fixed hours—600, 1200, 1800 and 2400 hours Greenwich Mean Time—trained observers at the various stations around the country and around the world take a whole series of measurements: air pressure, air temperature, cloud cover, wind speed and direction, visibility, moisture in the atmosphere near the ground and amount of precipitation since the last observation.

All the data go off in a succinct international code—by telephone, Teletype or radio—to a national or regional center that transmits it to other centers from which in turn it receives transmissions.

Nor is that all.

Twice a day, and sometimes more often, instruments called radiosondes are sent up, carried by hydrogen- or helium-filled balloons. As they climb at 1,000 feet a minute, the radiosonde sensors measure pressure, temperature and moisture, and their tiny radios send back the data until the balloons burst at altitudes of 60,000 to 70,000 feet. If a reflector made from a light metal mesh is attached to a balloon, radar can be used to track the radiosonde's drift, thus providing information about wind speed and direction.

Radar is used for other purposes as well. Radar employs radio waves of very short wavelength which travel straight and with the speed of light. When they strike an object in their path, they are reflected and the returning signals can be picked up by a receiving antenna. The time between emission and return indicates distance.

Because the waves are reflected by raindrops and hail, radar can locate storms. Taken at intervals, radar pictures can show the speed and direction of a storm, with the intensity of the light on the radar screen indicating the density of the precipitation.

Satellites orbiting the earth are in increasing use. They can measure infrared radiation flowing up through the atmosphere and this information can provide data on temperature and moisture content at various heights.

Other satellites of the geostationary type, which stay "parked" at great height over one area of the earth, can photograph clouds and provide data about air currents that move the clouds.

All this information would be of little value unless put into a form suitable for study. Mapping does that.

This is a basic tool in forecasting. A kind of atmospheric survey allowing visualization of weather conditions over large areas, the surface map is known technically as a synoptic chart—from the Greek *synoptikos*, meaning a general view or summary.

On the map, the location of each reporting station is indicated by a small circle, and, next to it, an identifying number. Information from the station is then marked in and around the circle.

The amount of shading in the circle shows the proportion of sky covered by clouds. Wind direction and speed are shown by an arrow drawn toward the circle. Barbs or feathers on the arrow indicate the wind speed, with a short feather denoting 5 knots, a long one 10 knots, a long and short 15 knots, etc.

Air temperature is shown along with dew-point temperature which is the temperature to which the air would have to be cooled to start condensation of its water vapor.

And air pressure is shown. Because the pressure decreases with height and a station on a mountain or hill would always measure lower pressure than another station at sea level, all stations calculate and report only an equivalent sea-level pressure. The pressure is measured in millibars (mbs) and the average sea-level value is 1,013.2 millibars. Typically, low pressure systems have values less than 1,000 millibars while highs have pressures greater than 1,015 millibars.

When all pressure values are plotted on the map, isobars can be drawn. These are lines that join places with equal pressure. Examining these lines, it becomes possible to note any highs and lows. If pressure increases toward the center of an area, it is labeled *H* for high; if pressure decreases, it is labeled *L* for low.

Look at a weather map and you'll notice from the wind arrows that the winds blow clockwise and slightly outward around a high and counterclockwise and inward around a low. If there are many isobars, that means a steep pressure gradient. You can consider isobars much like contour lines on a geographical map, each line

INTERNATIONAL STATION MODEL

C_M -MIDDLE CLOUD TYPE
C_H -HIGH CLOUD TYPE
$T_d T_d$-DEW POINT
a   -PRESSURE TENDENCY*
pp -CHANGE IN PRESSURE*
RR -AMOUNT OF PRECIPITATION (PAST 6 HOURS)
R_T -TIME PRECIPITATION BEGAN OR ENDED
N   -TOTAL SKY COVERAGE
dd -WIND DIRECTION

ff   -WIND SPEED
VV -VISIBILITY
ww -PRESENT WEATHER
W   -PAST WEATHER (3 TO 6 HOURS)
PPP-MSL PRESSURE
TT -TEMPERATURE
N_h -SKY COVERAGE OF "h"
C_L -LOW CLOUD TYPE
h   -HEIGHT OF LOW CLOUDS

* TENDENCY OR CHANGE OVER THE PAST 3 HOURS

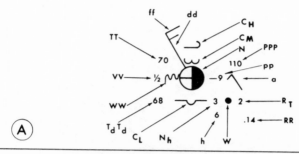

(A)

FACSIMILE STATION MODEL

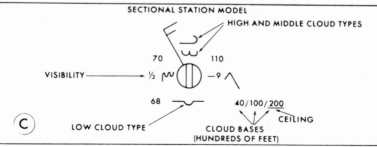

(B)

SECTIONAL STATION MODEL

(C)

aavn 221

Surface weather map station models.

81

joining points with equal altitude. The steeper the hill, the more contour lines. If water is flowing down that hill, the steeper the hill is, the faster the water will flow. Similarly, air will flow more strongly when there is a steep pressure gradient. So, the more isobars you see on the weather map, the stronger the wind; and

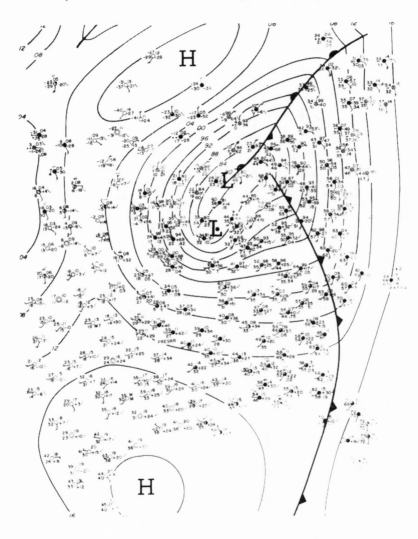

Surface weather chart, 1200 GMT, March 5, 1964.

the fewer, the more widely spaced the isobars, the more likely it is that the winds will be light or variable.

When you see an *H* on a weather map, you can interpret it as a mass of air moving clockwise, heavier than adjacent air, and heaviest at its center, with the contour lines indicating the pressure gradations. And you can visualize the air in a high sinking and, at the surface of the earth, flowing outward across the isobars, away from the center, and moving clockwise around the high.

When you see an *L*, you can view the low-pressure area as air moving counterclockwise, lighter than adjacent air, and lightest in the center, again with the contour lines indicating the pressure gradations. And in a low, the air will be flowing inward across the isobars and moving counterclockwise around the center.

Also drawn on the map are fronts and air masses. The weather forecaster analyzes the data on the plotted map. Where are the cold temperatures? The wind shifts? The regions of lowest pressure? By carefully connecting on the weather map the points of equal pressure (isobars) and of similar temperatures (isotherms), the boundaries (fronts) between the different air masses are established.

A cold front, the forward edge of an advancing cold air mass, is shown as a black line with small triangles having their points facing in the direction of the front's movement.

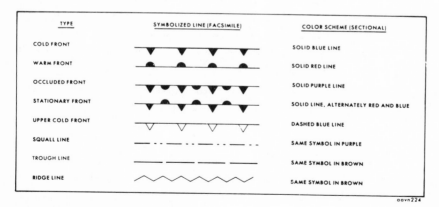

TYPE	SYMBOLIZED LINE (FACSIMILE)	COLOR SCHEME (SECTIONAL)
COLD FRONT		SOLID BLUE LINE
WARM FRONT		SOLID RED LINE
OCCLUDED FRONT		SOLID PURPLE LINE
STATIONARY FRONT		SOLID LINE, ALTERNATELY RED AND BLUE
UPPER COLD FRONT		DASHED BLUE LINE
SQUALL LINE		SAME SYMBOL IN PURPLE
TROUGH LINE		SAME SYMBOL IN BROWN
RIDGE LINE		SAME SYMBOL IN BROWN

aavn224

Weather map analysis symbols.

A warm front, the leading edge of an advancing warm air mass, is also shown as a black line—but now attached to the line in the direction of the front's movement are small black half-circles.

A stationary front which separates two air masses that are showing little movement appears on the map as a line with alternate triangles and semicircles.

To place a front, weather forecasters can make use of information about clouds, precipitation, pressure, and more. A well-developed front is usually marked by an area of clouds and precipitation more or less parallel to it. With a cold front, there will usually be a thin band of cumuliform clouds; with a warm front, there will be a succession of clouds from cirrus to stratus. There also is usually a change in wind direction at a front, and the isobars change direction abruptly. But it takes experience and skill to interpret all this.

UPPER-AIR MAPS

Having a weather map is fine. But, in itself, the map predicts nothing. It is a record of things as they are at a given time. What the forecaster has to do is to produce a similar, though not as detailed, map of conditions to be expected for a future time.

That means considering how far the systems shown on the synoptic map will move over the forecast period and how they may change as they move. And to add to the complexity, any changes that do occur can also influence the rate of motion.

At this point, information from upper-air soundings is needed because conditions up high can influence the direction and speed of motion of systems down near the surface of the earth. It's helpful to know what is going on right up into the stratosphere—wind direction and speed, pressure, temperature, humidity.

As one example of the importance of upper-air soundings, low pressure systems at the surface can be expected to be steered in the direction of upper-air flow and, moreover, their typical forward speed is about 30 percent of the wind speed at about 18,000 feet.

The data from upper-air analysis come from radiosonde recordings of pressure, temperature and humidity at various elevations. Balloon observations give wind direction and force at various elevations.

Upper-air maps look simpler than surface maps, with fewer highs and lows. But there are seven of them—for different elevations—to be considered along with the surface map.

This is where the computer comes in.

THE COMPUTER

The computer can be programmed so that, given the information needed to draw a synoptic surface map, it does the drawing. And given the data from the upper air, it proceeds to plot contours, temperatures and other data for every level.

And the computer can do more.

It can be programmed—given the proper mathematical equations—to go on, using all of the maps, to produce a forecast chart. In fact, it can print maps of conditions for the next six, twelve or twenty-four hours at the surface of the earth and at the upper air levels.

The computer-produced weather charts can be placed in an electric scanning machine and the pictures are then automatically transmitted and reproduced on sensitized paper at receiving weather stations all over the country.

Once the charts are received from the national center, regional forecast centers can add necessary local data. Similar pressure systems or fronts do not necessarily produce the same weather in all locations. It's to be expected, for example, that a front that has to move over hills or mountains will, because of the added lift for its moist air, produce more rain than a front moving over flat country. Cold air that has to cross a large lake may pick up moisture and produce snow showers on the downwind shore.

Your local newspaper and TV station may carry such maps along with official national, regional and local forecasts.

*   *   *

These are some of the tools, then, that make forecasting increasingly scientific and objective. The chances are that on a TV weather broadcast, the official reports will be paraphrased and interpreted. From time to time, though, the TV forecaster, out of his own experience, may issue a forecast differing from the official one, proving once again that there is still room for art—or, if you will, informed hunches—in making predictions.

7

Doing It Yourself

IS IT POSSIBLE for you to do your own forecasting? That is, without access to the weather charts and data available to the professional meteorologist? Of course. People do it all the time and have been doing it since long before there were weather services—often quite well, too, within certain limits.

Farmers, sailors and others, through shrewd observations and extended experience, developed sayings and figured out signs and portents, some of which have proved to be well-based.

For one thing, you can forecast in connection with official weather reports and those you see on TV, refining the predictions so they may apply better to the area where you live.

This is particularly useful if you are into boating or fishing. By learning a little about the weather and the forecaster's limitations you can keep out of trouble when on the water.

One of my viewers wrote to complain that he had been caught in a sudden squall that was entirely unexpected. The forecast, he said, called for brisk northwest winds and nothing more. Here was a case where the forecast was on target except for the timing.

A cold front was expected to move through the area during the early morning hours to be followed by a clearing trend and northwest winds. Ahead of the approaching cold front the winds were southerly, skies cloudy, with scattered showers and thundershowers.

Our boatman, depending on the previous night's TV weather forecast, and without getting an update on the position of the front early the next morning, went fishing. Had he checked, it should have been obvious to him that the expected passage of the cold front had been delayed. The timing was off! The barometer was falling, and he was out there on the bay, under gray skies, in southerly winds that were picking up speed.

A weather-wise fisherman would have waited. The front came through at noon, six hours late, triggering off the squalls that caught our unhappy viewer by surprise. After the squalls the weather did clear rapidly, and for the remainder of the day the dry northwest winds brought good fishing weather. An appreciation of the whole weather picture is important.

You must take into consideration the question of timing and listen carefully to the description offered by your weather person. A good forecaster will try to point up the limitations of the prediction, and from there on you are on your own.

If your hapless media meteorologist is in a dilemma and offers the following: "A cloudy, cool, drizzly day coming up—but if the warm front that is just south of us moves in we'll break out into southerly winds and warm sunshine," you know it's a toss up! You'll just have to wait and see if his first hunch is right, but if the winds do shift and it turns out to be a great day, at least it shouldn't come as a surprise.

Remember, too, when you employ the weather maps in your newspaper, to check the time of the map. It's generally thirty hours old already. But look for the positions of the storms (lows) and fair weather systems (highs). Then see where your TV weather map has them positioned. If you keep tabs on the weather this way you'll be able to listen to the weather forecasts in the morning and relate them in your mind to what you've seen the previous evening.

Say yesterday's newspaper map showed a high moving your way. Is that big high going to reach your area? Last night the hippy-dippy weatherman had it moving closer. And today the barometric pressure is high and rising, and the morning DJ said the official forecast was for sun today and tomorrow. But you know better! Sunny today—tomorrow—*and* into the weekend. Time to plan picnics or painting the house exterior. That's a huge high pressure system and they generally last for more than a day or two!

The various clues offered by TV weather segments—the wind,

the barometer reading, the relative humidity—are there for you to use.

WIND-BAROMETER INDICATIONS

Wind is weather on the move. If you know wind direction and speed, you have a good start toward forecasting local weather. Look at barometric pressure changes, too, in conjunction with what the wind is doing, and you're likely to do pretty well.

Taken together, wind direction and pressure change give you an indication of moving high- and low-pressure areas.

Generally, anywhere in the United States, it is highly likely that when the wind sets in from between south and southeast and the barometer drops steadily, a storm is approaching from the west or northwest and its center will pass near or north of you within twelve to twenty-four hours, with wind shifting to northwest by way of south and southwest.

It's also highly likely that when the wind sets in from points between east and northeast and the barometer drops steadily, a storm is approaching from south or southwest and its center will pass near or to the south of you within twelve to twenty-four hours, with winds shifting to northwest by way of north.

You can generally expect, too, anywhere in the United States, that the rate and amount of fall of the barometer indicate the speed of the storm's approach and how intense it is likely to be.

And, as a general rule, you can expect foul weather to follow when winds are from the east quadrants and the barometer is falling, while clearing and fair weather will follow winds shifting from the west quadrants.

It's important to remember that a single reading of a barometer accomplishes little. It is the kind and rate of change of pressure that count.

An aneroid barometer contains a bellowslike, airtight metal chamber from which most of the air has been evacuated. As outside atmospheric pressure rises and falls, the chamber contracts or expands. This reaction is indicated by the needle on the face of the barometer. Most home barometer scales are in inches—a reference to the standard mercury barometer in which measurements of air pressure are taken by observing the height of a column of mercury in a glass tube. For instance, the 14.7 pounds of air pressure per square inch average at sea level works out to be about 29.92 inches as read on a barometer.

Most aneroids are set by turning a setscrew located on the back of the instrument. It may be reached with a small screwdriver through an opening on the back of the barometer case. Use the sea-level pressure as broadcast by your local radio station, by National Weather Service VHF-FM (162.55 Megahertz, in the New York area), or as reported by phone when you call your local weather number, and turn the setscrew to obtain this reading.

Many aneroids are equipped with an additional pointer which may be turned by a knob to a position directly over the needle indicating the pressure. The next time a pressure reading is taken, it is then possible to see immediately the change in pressure by noting the difference between the readings of the pointer and pressure indicator needles.

The barometer may be kept in any room in the house not subject to sudden temperature change. There is no reason to locate the instrument outdoors since the pressure is about the same in an ordinary room as it is outside.

As you may know, although barometers fluctuate with the passage of highs and lows, the range of change is not very great. There is some variation in range from one area to another but in any one place it is not uncommon for the difference between the highest and lowest barometer readings to be no more than about 2½ inches, or about 85 millibars—a difference in pressure of little more than a pound per square inch.

When it comes to wind direction and speed, ou can, of course, use equipment to measure both—a weather vane for direction and an anemometer for speed. But you can also do pretty well without instruments.

For direction, you can, for example, observe a flag or just go outside and feel where the wind is coming from.

Not at all incidentally, you can go out, put your back to the wind, and judge the location of a high or low. Because winds blow clockwise around a high and counterclockwise around a low in our hemisphere, when you stand with your back to the wind, lower pressure will be to your left and higher pressure to your right.

As for wind speed, one way to judge it is by observing a flag. The flag will hang limply in a calm. At about 10 miles an hour it will wave out, about one-third from the perpendicular staff; at about 20 miles per hour, it will be about two-thirds up; and at 30 miles per hour or more, it will wave straight out in a fully horizontal position.

You can also judge wind speed by its effects on other common objects, using the following table:

Wind (mph)	Effects	Official Designation
Under 1	Smoke rises vertically. No perceptible motion of anything.	Light
1–3	Wind direction shown by smoke drift, but not by weather vane.	Light
4–7	Leaves rustle slightly. Wind felt on face. Weather vane moves.	Light
8–12	Leaves and twigs move; dust raised from ground; wind extends light flag.	Gentle
13–18	Small branches move. Dust and paper move.	Moderate
19–24	Small trees in leaf begin to sway. Large branches move. Dust clouds.	Fresh
25–31	Large branches move continuously. Wind whistles. Difficult to use umbrella.	Strong
32–38	Whole trees sway. Walking into wind difficult.	Strong
39–46	Twigs break off trees. Cars veer.	Gale
47–54	Slight structural damage; chimney pots, roof slates may blow away.	Gale
55–63	Trees uprooted. Much structural damage.	Whole Gale
64–72	Widespread damage.	Whole Gale
73 or more	Widespread damage.	Hurricane

91

Here are some general hints about the kind of weather you might expect with different combinations of wind direction and barometric readings and trends.

Wind direction	Barometer reduced to sea level (in inches)	Character of weather indicated
SW to NW	30.10 to 30.20 and steady	Fair, with slight temperature changes for 1 to 2 days.
SW to NW	30.10 to 30.20 and rising rapidly . .	Fair, followed within 2 days by rain.
SW to NW	30.20 and above and stationary	Continued fair, with no decided temperature change.
SW to NW	30.20 and above and falling slowly .	Slowly rising temperature and fair for 2 days.
S to SE	30.10 to 30.20 and falling slowly . . .	Rain within 24 hours.
S to SE	30.10 to 30.20 and falling rapidly . .	Wind increasing in force, with rain within 12 to 24 hours.
SE to NE	30.10 to 30.20 and falling slowly . . .	Rain in 12 to 18 hours.
SE to NE	30.10 to 30.20 and falling rapidly . .	Increasing wind, and rain within 12 hours.
E to NE	30.10 and above and falling slowly .	In summer, with light winds, rain may not fall for several days. In winter, rain within 24 hours.
E to NE	30.10 and above and falling rapidly	In summer, rain probable within 12 to 24 hours. In winter, rain or snow, with increasing winds, will often set in when the barometer begins to fall and the wind sets in from the NE.
SE to NE	30.00 or below and falling slowly . .	Rain will continue 1 to 2 days.
SE to NE	30.00 or below and falling rapidly .	Rain, with high wind, followed, within 36 hours by clearing, and in winter by colder temperatures.
S to SW	30.00 or below and rising slowly . .	Clearing within a few hours, and fair for several days.
S to E	29.80 or below and falling rapidly .	Severe storm imminent, followed within 24 hours by clearing, and in winter by colder temperatures.
E to N	29.80 or below and falling rapidly .	Severe northeast gale and heavy precipitation; in winter, heavy snow, followed by a cold wave.
Going to W	29.80 or below and rising rapidly . .	Clearing and colder.

Temperature and humidity, considered together, can often provide an indication of weather change.

There is always water in the form of invisible vapor in the air, and it is when the vapor condenses that we get much of our weather phenomena, rain, snow, clouds, fog and dew, among them.

Relative humidity is defined as the ratio of the amount of water vapor actually present in the air to the maximum amount the air is capable of holding at a particular temperature. It is expressed in terms of percentage. The amount of gaseous water vapor that the air can hold varies with the temperature. At higher temperatures air can hold more water vapor than it can hold at lower temperatures. Therefore as the temperature increases, if there is no change in the total amount of water vapor from some outside source, the relative humidity will decrease. Conversely as the temperature decreases, relative humidity will increase. This is why most areas show a maximum relative humidity about sunrise, when temperatures are normally coolest, and why afternoon relative humidities are usually lower than night and morning ones. It is also the reason that indoor relative humidities are frequently low during winter months. The outside air is cold, and even when saturated, it holds relatively small amounts of water vapor. When the air temperature is raised by the inside heating system, the relative humidity shows a large decrease. Roughly an increase in temperature of 20 degrees Fahrenheit will reduce the relative humidity by one-half; a decrease in temperature of 20 degrees Fahrenheit will double the relative humidity. Thus outside air with a relative humidity of 90 percent at 20 degrees Fahrenheit will have approximately a 45 percent relative humidity at 40 degrees Fahrenheit, a 22 percent relative humidity at 60 degrees Fahrenheit and an eleven percent relative humidity at 80 degrees Fahrenheit.

Relative humidity can be determined in various ways. One method employs a hair hygrometer, which registers changes in the length of human hair resulting from changes in atmospheric moisture. Another depends on changes in conductivity of certain salt solutions with changes in relative humidity.

Meteorologists and some earnest amateur forecasters most often use a sling psychrometer—a hygrometer consisting essentially of two thermometers, one ordinary dry-bulb thermometer, the other called a wet-bulb thermometer. The latter is a conventional thermometer but its bulb is covered with a muslin bag from which a wick of absorbent material dips into a vessel containing water. Water passes up the wick to the muslin, evaporates, and in so doing cools the thermometer bulb. As a result, the wet-bulb thermometer usually registers a lower temperature than the dry bulb.

But when the air is saturated, vapor can't be evaporated from the wet bulb; the air can hold no more. Under those circumstances, both wet- and dry-bulb thermometers give the same reading, meaning the relative humidity is 100 percent.

Most of the time, the two thermometers will give different readings. When it's very humid, the difference will be small; when humidity is less, the difference between readings will be larger. From the differences can be determined, with the help of humidity tables, the relative humidity and the dew point, the temperature at which dew would form.

We said that temperature and humidity, considered together, can often provide an indication of weather change. Let humidity increase while temperature remains the same—or let the opposite happen and while humidity holds steady the temperature drops sharply—and you can figure that rain is possible.

Of course there are people who become interested in the weather to the extent that they purchase weather equipment to provide accurate readings. But you don't have to set up a weather station. Just an awareness and close attention will soon make you weather wise.

Fog generally forms when the air is saturated with water vapor, or when the air is cooled down to the point when the dewpoint temperature is reached. In either case the relative humidity

Table of Relative Humidity
Temperature of air, dry-bulb thermometer, Fahrenheit

Difference between wet-bulb and dry-bulb readings	30°	40°	50°	60°	70°	80°	90°	100°
1	90	92	93	94	95	96	96	97
2	79	84	87	89	90	92	92	93
3	68	76	80	84	86	87	88	90
4	58	68	74	78	81	83	85	86
6	38	52	61	68	72	75	78	80
8	18	37	49	58	64	68	71	71
10		22	37	48	55	61	65	68
12		8	26	39	48	55	61	65
			68					
14			16	30	40	47	53	57
16			5	21	33	41	47	51
18			13	26	35	41	47	
20				5	19	29	36	42
22					12	23	32	37
24					6	18	26	33

approaches 100 percent. There are various kinds of fog that can form. There's the advection type, when moisture-laden air moves inland over cool ground and is chilled so that fog forms. Sea fog is another kind of fog that forms over the water and then drifts inland. Another common form of fog is radiation fog. At night the ground cools off rapidly under clear skies so that toward daybreak the temperature and dewpoint approach each other with the relative humidity climbing to near 100 percent. Fog may then form that is burned off with the early warming sun.

They've been called signposts in the sky. And observing them can tell you much about the state of the weather, since they help to indicate what the atmosphere is doing by providing visible evidence of air movement and water content.

Clouds are simply, if one can use that word, visible aggregations of tiny water and/or ice particles, differing from fog only in that the base of fog must be within 50 feet of the ground. Clouds form because of condensation of water vapor in air that is rising and cooling.

Generally, condensation takes place when the air is very nearly saturated and when plenty of condensation nuclei are present. Water vapor tends to condense onto such nuclei, which are minute solid particles, including dust, smoke, ash from fires, sea salt from spray, all suspended in the atmosphere.

These nuclei, of which the atmosphere is virtually full, have a vital role. Tests have shown that clean air can have a relative humidity as high as 400 percent without condensation occurring; the nuclei allow condensation at relative humidities of about 100 percent.

It was Luke Howard, an English chemist and amateur meteorologist who, in 1803, devised a basic cloud classification system that has become the foundation for the system adopted much later by the World Meteorological Organization.

Howard divided clouds into three basic classes according to their appearance: cirrus, from the Latin word for "curl of hair"; stratus, from that meaning "spread out"; and cumulus, from the Latin for "pile."

Elevation of clouds is another means of classification. There are high (16,500-45,000 ft.), middle (6,500-16,500 ft.), and low clouds (surface to 6,500 ft.). The high clouds are cirrus, cirrocumulus, cirrostratus and the tops of cumulonimbus. The middle: altocumulus, altostratus, nimbostratus and portions of cumulus and cumulonimbus. The lows: stratocumulus, stratus, cumulus, and the bases of cumulonimbus.

As you can see, in addition to the three basic types of clouds classified by appearance, variations occur that have been given combined names such as cirrocumulus and cirrostratus.

High clouds usually occur at elevations of 4 to 8 miles, middle at heights of 1 to 4 miles, and low at elevations up to about 6,500 feet.

It's through clouds, of course, that water is returned to earth and, in fact, distributed around the earth, after being picked up at an estimated rate of about 80,000 cubic miles evaporated from the oceans and 15,000 cubic miles from rivers, lakes, wet earth and vegetation. Worldwide, the average annual rainfall is about 40 inches, and a fourth of it, some 24,000 cubic miles, drops back on land, enough on average to provide some 22,000 gallons of rainwater for each individual on earth, man, woman and child. Still, distribution is hardly even. The rainiest spot on earth, reputed to be Mount Waialeale in Hawaii, gets about 470 inches a year; and there are desert areas getting less than an inch annually. In the continental United States, the range is from some 1.7 inches in Death Valley in California to 140 or more inches in Pacific Northwest coastal areas.

If you're interested in forecasting, you'll find it useful, indeed, to know a few basic facts about clouds.

Cirrus clouds, made up of various sizes of ice particles, have a feathery, delicate or fibrous appearance. They can take varied forms such as tufts, streaks and plumes, often spread over all of the sky yet not blotting out sun or moon. Long, well-defined wisps of cirrus are sometimes called mares' tails.

Cirrocumulus, made up of supercooled water droplets or small ice crystals or mixtures of both, are small and delicate cloudlets that arrange themselves in lines and ripples. The term "mackerel sky" is often used to describe them. Cirrostratus clouds, fibrous, white or gray, often cover the sky giving it a milky appearance and often produce halo phenomena. They don't blot out the sun. A cirrostratus cloud often thickens to form, first, an altostratus, and then later, a nimbostratus cloud, and therefore can be an indication of oncoming rain.

Cumulonimbus top is a dense cirrus cloud made up of super-cooled water droplets and ice crystals, mostly the latter. It spreads out from a cumulonimbus top (see cumulonimbus, below) to form an anvil-shaped extension.

Those are the high clouds. Now the middle:

Altocumulus consists of white or gray layers or patches of solid cloud, sometimes occurring in a regular pattern of lines or waves, producing a mackerel sky.

Altostratus is a stratified veil of clouds ranging from gray to dark blue, often covering the entire sky. If altostratus is thin, you can usually see the sun or moon through it but without the halo phenomenon that may appear with cirrostratus. Precipitation often falls from altostratus and is usually light and fairly continuous.

Nimbostratus is a gray or dark massive cloud formation, often obscuring the sun completely, and is made up of suspended water droplets and falling raindrops or snowflakes. Often nimbostratus produces continuous rain, snow or sleet, unaccompanied by lightning, thunder or hail.

Cumulus and cumulonimbus found at the middle level are extensions of the low clouds, ascending to middle-cloud height range (see below).

And finally, to the low clouds:

Stratocumulus occurs in gray or whitish layers or patches of flakes or globular masses, with rolls or rounded portions, and is generally made up of small water droplets, sometimes accompanied by larger drops, soft hail, and, rarely, snowflakes.

Stratus is a uniform, somewhat foglike layer of cloud (but not resting on or near the ground) made up of minute water droplets or, sometimes, when temperature is low enough, partly of ice crystals. It is often formed by the lifting of lower layers of a fog bank. It usually does not produce precipitation.

Cumulus is a thick cloud with a lot of vertical development. Its base is nearly horizontal while its upper portions often look like a cauliflower head. It's made up of small water droplets. Larger droplets often form within the cloud and fall from the base as rain.

Cumulonimbus is a heavy mass of cloud often extending from

Right, my son Storm is on WABC-TV. We face each other nightly on the six and eleven PM news. We also compete against each other in covering the science field. *Courtesy ABC*

Below left, Pamela Diane Field is a science reporter and my co-host on *Health Field,* a regularly syndicated program. *Courtesy NBC*

Below right, Allison Carol Field is now a full-time reporter on WNBC-TV. She also fills in on the weather when our regular weekend weatherman is away. *Courtesy NBC*

Above left, Willard Scott is riding high on NBC's "Today." He's a professional radio and television announcer who found his niche as a weatherman. *Courtesy NBC*; *Right*, a schoolteacher finds fame and fortune as a TV weatherman—Dr. George Fischbeck. *Courtesy WABC-TV*; *Below*, Washington, D.C.'s meteorologist, Bob Ryan, made the jump from research to performing with great success. *Courtesy WRC-TV*

In response to Johnny Carson's gibes about Dr. Frank Field, "NBC's crack meteorologist," I would dub Johnny's head onto photos each night as part of my weathercast. *Courtesy NBC*

Above, on my first appearance on the "Tonight Show" I was caught in a snowstorm after predicting sunshine. *Courtesy NBC*; *Below*, one night Johnny Carson did a surprise walk-on and proceeded to demolish the late weather report, much to the anger of the NBC News brass. *Courtesy NBC*

This wirephoto spread Winston's fame world-wide and got this meteorologist in hot weather. *Courtesy NBC*

When Winston was attached by NBC management, he received moral support from Tom Snyder, our anchorman, and me. *Courtesy NBC*

Left, As NBC Science Editor, I went on location in Israel's South Haibar game preserve to examine firsthand the ways in which Israel is working to restore the animals of the Bible to the land of the Bible. The "Eternal Light" broadcast, "All Thy Creatures," was televised this year. *Courtesy NBC; Below,* This was the first regular use of global weather satellite photos on NBC network television during the United States space program in the 1960s. *Courtesy NBC*

SUBPOINT   .5N   75.5W

This full-disk image from the East Coast GOES (Geostationary Operational Environmental Satellite) on August 7, 1980, shows Hurricane Allen entering the Gulf of Mexico. A second hurricane, Howard, is in the Pacific south of Baja California. *National Oceanic and Atmospheric Administration*

03SE79 12A-2 01033 17971 DB5

04SE79 12A-2 01033 17941 DB5

Hurricane David battering the Atlantic states from Florida to New England, September 3-6, 1979. The eye of the storm moved over land, maintaining its strength, which is unusual for such tropical storms. *National Oceanic and Atmospheric Administration*

05SE79 12A-2 01034 17961 DB5

06SE79 12A-2 01031 17981 DB5

lower atmospheric levels (2,000–4,000 feet) through middle-cloud altitudes up into the high-cloud region, with summits taking the shape of mountains or towers. Cumulonimbus is accompanied by lightning and thunder as well as rain, snow or hail.

You can often forecast the coming of a warm or cold front by examining the cloud formations.

Along a warm front, you'll recall, warm air is riding up over colder denser air along a sloping surface, cooling as it rises, and, with the cooling, the water vapor condenses into clouds.

About 500 miles ahead of the front, at its leading edge where the sloping surface is about 5 miles high, the coming of the front will be heralded by high, thin wisps of cirrus clouds.

As the front advances, clouds become progressively lower. The cirrus are replaced first by high cirrostratus, then by the altostratus or altocumulus clouds of middle altitudes, at which point rain may fall, and the air becomes warmer and more humid. Finally, when the trailing portion of the front approaches at ground level, there will be heavy layers of low nimbostratus clouds with steady rain. The next day, the sun is likely to break through the clouds, and there follows a spell of warm weather.

A cold front produces a different sequence of clouds and weather. The mass of cold air bears down and plows under the warmer air, tossing it upward. Often, there is little or no warning: The cold is suddenly there at ground level. As the front moves in and the warm air is thrown up, there are often towering cumulonimbus clouds, dark and threatening, and frequently producing violent thunderstorms or very heavy showers. Then the sky often clears as quickly. But the cold air sometimes is unstable, leading to development of cumulus and cumulonimbus clouds and cool, showery weather.

The cumulus clouds of fair weather usually result from convection induced by ground-level heating. Tall cumulus, sometimes tens of thousands of feet high, may develop over warm bodies of land or water, and smaller ones often develop over such man-made warm spots as metropolitan pavements and plowed rural fields.

Although good weather usually is accompanied only by fluffy

cumulus clouds, on a hot and muggy day even without an oncoming cold front, a vast cumulonimbus cloud, as much as 6 miles high and that much or more in diameter, containing hundreds of thousands of tons of water, can form. Within that cloud, with its warm air at bottom and cold at top, violent up- and downdrafts take place and rain pours down.

What actually causes precipitation? Why is it sometimes in the form of rain, sometimes snow, sometimes hail?

There appear to be two processes involved, at least so far as rain is concerned. One holds true for the tropics, the other for the temperate zones, where we live.

In hot latitudes, convection currents are quite strong and the air turbulent. Cloud droplets churn, have greater chances of collision, and tropical air tends to be full of salt crystals that provide nuclei for many droplets which then become somewhat larger and faster falling than average. These collide with smaller droplets on their way downward, form still larger droplets, which fall still faster and pick up more small droplets, until they finally become so large and heavy that they fall out of the cloud as rain.

In our latitudes, however, most rain may actually begin as particles of ice or snow. That's the theory first proposed in 1935 by Tor Bergeron, a Swedish meteorologist, since elaborated on by others, and now generally accepted.

According to the theory, at least the tops of all clouds from which appreciable rain falls are at temperatures below freezing and consist of both ice crystals and supercooled water droplets. The supercooled droplets are drops of water that were condensed to liquid form at temperatures above freezing and remained liquid even after being cooled to temperatures below freezing. There is no full explanation for this phenomenon but it is known to happen often in the atmosphere.

It's also known that when ice crystals and supercooled droplets exist together in a cloud, there will be evaporation from the droplets and condensation on the crystals which thus grow to sufficient size to fall.

As they fall, the crystals collect more droplet condensation. If temperatures are cold all the way to the ground, they may hit ground as crystal aggregations, in the form of snowflakes. Or, after leaving the cloud as snowflakes, they may encounter warm air, melt, and reach earth as rain. Sometimes, after melting on encountering one layer of warm air several thousand feet above earth, they may strike colder air nearer the earth, refreeze, and arrive as sleet, which consists of ice pellets.

Hail, still another kind of precipitation, falls only from cumulonimbus. Such clouds have updrafts of warm, moist air in which condensation often begins as rain, but the drops are carried upward by the rapidly rising warm currents. When they reach cloud regions where the temperature is below freezing, they congeal and acquire an ice coating. Then they descend to lower portions of the cloud where they get a coating of water which freezes around the cold centers, and they then are carried up again into the ice region of the cloud. In these up and down journeys, they grow rapidly. According to a newer theory, such journeys may or may not be necessary, and hailstones may be formed just by falling through subfreezing air layers and acquiring supercooled water drops with which they collide.

Hailstones sometimes acquire alternate layers of clear and opaque ice and may reach large size before falling to the ground. Stones as large as marbles are common and sometimes the stones are of considerably greater size. At Potter, Nebraska, on July 6, 1928, one stone was 5 inches in diameter and weighed 1½ pounds.

If you keep track of clouds and what happens to them, often also taking into account winds, it's possible to arrive at some predictions for the short term.

Start with the high clouds—cirrus, cirrostratus and cirrocumulus. As long as they stay thin, are present only in isolated patches, and winds are from a westerly quadrant, you can expect fair weather with little change.

But if they thicken or join together, gradually lower, with thicker clouds forming underneath, and they move from south or southwest while winds at the surface are veering from the east, precipitation within twelve to twenty-four hours is probable.

When it comes to middle clouds—altocumulus and altostratus—if patches of altocumulus appear to fuse, becoming sheetlike or developing into layers, and their movement is from the south or southwest while surface winds are veering and easterly, precipitation within six to twelve hours with southeast winds is probable, followed by partly cloudy and warmer weather with southwest winds.

If you see altocumulus moving rapidly from west or northwest while surface winds are freshening from south or southwest, there may be hard, brief precipitation (sometimes, squall possibilities), followed by clearing and cooling temperatures.

If you find altocumulus patches appearing on a hot summer midmorning, with castlelike tufts, there may be strong local gusts and thunderstorms during the day.

But as long as altocumulus masses stay isolated and you see open sky between them and they are moving in the same general direction as winds at the surface, it's likely that weather should stay fair.

If you see altostratus layers darkening and lowering from the south or southwest, with the sun becoming faint, and surface winds veering from northeast, east and southeast, a warm front is due in a few hours, with poor weather and steady precipitation, followed by warmer southwest winds and partly cloudy skies.

As for low stratocumulus, stratus and nimbostratus clouds, as long as stratocumulus have a rounded appearance, with open sky between them, and they drift pretty much in the same direction as surface winds, weather should be unchanged for the next day or so.

But if they fuse, become lead gray, and form an overcast, expect unsettled and rainy weather.

If you see stratocumulus appearing as a long roll of cloud in the west or northwest, and surface winds are fresh or strong and from the south or southwest, you may be in for sudden squalls, shifting winds, hard but brief precipitation, then clearing and cooler weather.

When you find an overcast of thick stratocumulus beginning to show breaks, with cloud bases rising, and there are steady westerly winds, gradual clearing and cooler weather are likely.

When you see stratus in the morning and winds are light, it's likely that the sun will break up the cloud and there will be fair weather.

If you see a dark nimbostratus overcast following after lowering altostratus, and cloud movement is from the south or southwest while surface winds are from the southeast, steady precipitation is likely, and after several hours, warmer weather can be expected, with wind shifting from the south or southwest.

Cumulus or cumulonimbus clouds can have a lot of vertical development, with a lower level of about 1,500 feet and an upper level reaching cirrus heights.

If the cumulus do not develop until afternoon and are alone in the sky, they will generally be gone by end of day, leaving clear skies and fair weather ahead.

When, on a humid day, you find heavy cumulus forming late in the morning and building up higher during the afternoon, you can probably expect scattered showers late in the afternoon.

Look for scattered thundershowers in afternoon or evening, possibly with some lightning and hail, when cumulonimbus appear in isolated masses and move individually from south or southwest.

If you see a line of cumulonimbus or heavy cumulus turning up from the west, northwest or north, and winds are fresh to strong from south or southwest, expect quickly developing squally weather and hard thunderstorms followed by clearing skies and cooler, drier weather.

And when heavy cumulonimbus clouds develop very dark, almost black bases that look like they're building downward, the likelihood is that severe thunderstorms and hail will develop quickly, with tornadoes possible.

*   *   *

Put more generally, you can expect increased likelihood of bad weather if you see any of the following: Isolated high patches of clouds thickening, increasing and lowering; fast-moving clouds doing the same thing; clouds moving in confused fashion, from different directions at different times; clouds developing dark bases; middle clouds darkening the western horizon; heavy clouds building upward on a summer morning.

On the other hand, better weather is likely with any of the following: a steady decrease in the number of clouds, increasing breaks in the overcast produced by a stratiform cloud, fog gone before noon.

FOLKLORE

How much reliance can you place on all the multitudinous bits of folklore and proverbs about the weather?

Many are pure myth. Some turn out to have a scientific basis and can be helpful for predicting.

Let's start with some of the myths:

Flies bite more viciously before storms. This one dates back more than 2,000 years to Theophrastus, a pupil of Aristotle, whose *Book of Signs*, written about 300 B.C. described better than 200 portents of rain, wind and fair weather. Just not true. Flies can bite just as hard and often before fair weather.

If on Groundhog Day the groundhog, coming out of hibernation, sees its shadow, it will go back for six more weeks and winter will be prolonged. There is no connection whatever between a sunny February 2 and length of winter.

When squirrels store a lot of nuts, winter will be severe. When squirrels store a lot of nuts in fall, it's because there are a lot of nuts around to store.

In like a lion, out like a lamb. This usually refers to March. But there is no scientific case for believing that if March 1 is stormy, March 31 will be mild and pleasant.

When frogs croak, look for rain. Especially in Europe, exces-

104

sive croaking has been taken to be a sure sign of rain. Actually, the frog's croak has nothing to do with weather; it's a mating call.

And, while we're debunking, what about almanac weather forecasts?

The best almanacs have some value—but not for providing accurate weather forecasts. They provide, at best, climatological data for a region, that is, weather information about average conditions noted in that region over a period of many years. A reader can gain an idea of the kind of weather most often experienced in that area at a particular time of year. You couldn't expect an almanac to be right every time but it could be on or nearly on target some of the time, perhaps better than half the time.

Most almanacs, however, make broad, general predictions, often so worded that they will seem right to somebody somewhere at some time.

Dr. Robert Harnack is Professor of Meteorology at Rutgers University in New Jersey. He is also the Chairman of the Long-Range Forecasting Committee of the National Weather Association and this is his statement on long-range weather forecasts, including those found in almanacs:

> Reliable long-range forecasts for periods of time extending beyond five days from the present time have been difficult to achieve. The consensus among theoretical meteorologists is that *daily* predictions of weather beyond a few weeks in advance will always be no better than a random guess. It has not been demonstrated that daily forecasts, say of temperature or precipitation occurrence, can even be made with any reliability beyond five days at the present time. Again, in order to be considered reliable to any extent, a *set* of forecasts must be deemed correct in more instances than random guessing would achieve. Of course, even random guessing is bound to be correct some of the time, so the reliability of a forecaster or a forecast method cannot be assessed by verifying a selected small

105

number of forecasts and certainly not by verifying a single forecast.

Long-range predictions for the general character of temperature and precipitation conditions for a *period* of time such as for the next month or season have shown some, although small, amounts of reliability. Responsible meteorologists involved in long-range forecasting continue to try to improve the methods for making these generalized monthly and seasonal forecasts rather than the futile exercise of making specific daily forecasts beyond a week into the future.

Now let's consider some old adages that do have some basis in scientific fact:

Red sky in the morning, sailors take warning. Many other factors, of course, are involved in determining whether we'll have stormy weather, but at least it is a fact that a red sky indicates that some needed rain elements, such as dust and moisture, are present.

Halo around sun or moon, rain or snow soon. Not always, but sometimes—perhaps even more often than not. A halo indicates high cirrus clouds, which often precede an approaching warm front.

Mares' tails and mackerel scales make tall ships take in their sails. As we've seen earlier, mackerel sky is often used to describe cirrocumulus clouds and cirrus clouds are often called mares' tails. And such clouds often appear before an approaching warm front and precipitation along with veering winds.

Clear moon, frost soon. Possible. When the moon is clearly visible, the atmosphere may be largely free of clouds, and the earth can radiate away heat, cooling rapidly so that, given low enough temperatures, frost may form.

Rainbow in the morning, fair warning. Also possible. In the morning, with the sun in the east, a rainbow and its shower will be seen in the west. And since the west is where our weather generally comes from, there could be rain moving our way.

At night, of course, with the sun in the west, a rainbow will be seen in the east, which means the rain has passed and will keep moving to the east.

8

Special Weather/Special Forecasts

In the western Pacific they're called typhoons; near Australia, willy-willy's; and in the Indian Ocean, cyclones. They're known to us as hurricanes.

Whatever the name, they are whirling windstorms of tremendous power—nature's most destructive forces. The hurricane is huge in size. The danger it poses is not in the speed at which it moves. This storm may drift along at just a few miles per hour. The damage a hurricane creates is related to the winds that sweep around its center and the torrential rains and flooding these storms cause.

Almost everybody has experienced winds of 35, 40, perhaps even 50 miles an hour, which can blow off garbage-can lids, break twigs and do a bit of damage.

In a hurricane, winds reach velocities of 75 to 200 miles an hour, striking so forcefully that they can rip clothes off people's backs, and sweep not only cars off roads but trains off tracks.

As a hurricane nears land, it drives water ahead of it in the form of a storm surge, or "tidal wave," that can inundate lowlands along the coast with deep water. The storm surge is as likely to be responsible for deaths and damage as the winds or flooding due to heavy rain. In one of the nation's severest hurricanes, Camille in August 1969, 256 people died and property damage reached almost $1.5 billion.

With the approach of a hurricane, the barometer begins to fall, slowly at first then with increasing rapidity, and wind increases from breeze to hurricane force while clouds thicken from cirrus and cirrostratus to dense cumulonimbus, accompanied by thunder, lightning and heavy rain.

These conditions persist for hours, spreading destruction, until, suddenly, the eye of the hurricane arrives. Then wind dies down, rain stops, the sky clears partly or entirely, and barometric pressure stays steady at its lowest point.

The calm may last for half an hour or longer. And then comes the storm again, just as severe as before, but now the wind is from the opposite direction and pressure rises rapidly. While this continues, the wind gradually decreases until the storm is gone.

It has been estimated that the total energy in a hurricane of average fury may be the equivalent of forty hydrogen bombs exploded every second.

Origins. Tropical cyclones always originate in regions where sea surface temperature over a large area is high. They develop frequently only over very warm oceans when sea surface temperatures are 79 degrees Fahrenheit, or higher. Thus they originate in limited portions of all tropical oceans except the South Atlantic and the eastern part of the South Pacific where cold ocean currents keep surface-water temperature continuously below 79 degrees.

The most characteristic place of origin is in the doldrums about 10 to 20 degrees from the equator. These storms do not occur closer to the equator than about 5 degrees. The reason appears to be that the Coriolis force is needed for their formation (this is zero near the equator). And in the latitudes of the doldrums, they get that force along with the continuing supply of warm, moisture-laden air that goes into forming and maintaining them (hurricanes become weaker and soon lose their destructive intensity when they move over land).

A tropical cyclone may develop through several stages to maturity. In the formative stage, it is called a tropical depression, with winds up to 34 knots. It may continue to develop into a tropical storm, with winds of 35 to 64 knots. In the mature stage, when it attains hurricane stature, the winds reach 65 knots and often far more.

On average, according to records kept over a period of more than 40 years, the number of tropical storms developing annually in the North Atlantic is 9.5, with 5.5 attaining hurricane proportions, and of these an average of 1.8 reaching the coasts of the United States.

But averages don't mean very much. During that record-keeping period, up to five hurricanes struck us in one year while in six years not a single storm touched us.

Hurricanes are most common between June and November, usually peaking in September but they may also occur in May and December.

Once a hurricane forms, its usual movement in our hemisphere is toward the west and northwest until it reaches about 25 degrees latitude when it moves north and then northeastward. But there are many variations.

How do hurricanes form? There have been many theories. According to one of the most widely held, a typical Atlantic hurricane may have its origins when the steady flow of the easterlies, or trade winds, over warm seas is diverted northward by a low-pressure disturbance.

Because they are squeezed at the diversion curve, the winds may pile up on each other, and their warm air then may shoot upward as high as 40,000 feet, giving off heat and moisture before descending.

But something else is happening, too. Because of the earth's rotation, the rising column of warm air develops a twist and gradually assumes a cylindrical form so that it whirls about an inner core of relatively calm and still air. At the same time, more warm moist air picked up from the sea feeds the whirling column.

Tracking. Not too long ago—about twenty-five years—a hurricane that formed in the mid-Atlantic could move toward the coastal United States undetected. The first warning might come from one of the few ships crossing the Atlantic, if one inadvertently blundered into the path of the storm.

Today, hurricanes are still located by ships at sea, and by "hurricane-hunter" planes patrolling the Caribbean and Atlantic, but overwhelmingly thanks to satellite photos that show the unmistakable cloud pattern of a hurricane, there is little chance of one striking without warning.

In a hurricane, a wall of towering cumulonimbus forms around the eye and the clouds rise to great heights. Near the top the cloud droplets are replaced by ice crystals that are blown out from the storm center at high levels, forming a high deck of mares' tails, or cirrus clouds.

Once a hurricane is spotted and it begins to bear toward the United States mainland, planes are sent out several times a day by the National Hurricane Center in Miami. Crammed with instruments, the planes penetrate the storm area at different levels, going through the eye. They determine barometric pressure, temperature, humidity, wind direction and speed through the storm, trying to determine its force and the factors that may determine its path.

Penetrating the storm is no simple matter. It means enduring several minutes of very high winds and torrents of rain until the eye is reached—where the sun or stars may be out. But then the plane leaves the eye and has to take a repeat of the original buffeting.

The plane's findings may be summarized on your radio or TV weather broadcast. You will be told almost precisely where the center of the storm is, the pressure there, the wind speed and how far the winds reach out, and whether there are indications that the storm may intensify over the next twelve to twenty-four hours.

It is upper-air circulation that is critical in determining the path of the hurricane. And upper-air soundings by the planes are used by forecasters ashore to predict the forward movement speed and direction. Accuracy of these predictions, which are aided by computers, is high for periods of up to twelve hours when the upper-air currents provide a strong steering force. And once a hurricane arrives within radar range—roughly 250 miles from Miami—and other points as well—it can be kept under constant surveillance.

There are problems, however, when the steering forces are weak. Then the hurricane may move at different speeds or may be stationary for a time, or move one way and then another.

The role of the television weatherman is to keep the public up to date on the progress of the storm. The official warnings issued by the National Weather Service—such warnings as a storm watch (an indication that the hurricane could possibly affect your area) and storm warning (the storm is expected to affect the region)—must be adhered to. The competent TV weather reporter will not try to second-guess the official warnings.

Can anything be done to reduce the force of hurricanes? The U.S. Weather Bureau's Project Stormfury has been investigating methods.

A major hope lies in a chemical, silver iodide, which, when dropped into a hurricane, can sometimes stimulate cloud droplets to freeze. The hope is that if the technique can be refined and perfected, it may allow air currents in and out of the hurricane to be interrupted enough to dissipate the storm. But such weather control remains highly experimental.

TORNADOES

Of all the disturbances in the atmosphere, tornadoes are the most violent. Although short-lived and local, with rather limited paths, they can cause terrible destruction.

In January 1969, when a tornado moved across part of Mississippi, 32 people were killed and over 200 injured. In February 1971, midwest tornadoes killed 108 people and caused millions of dollars of damage.

A tornado is a violent circular whirlpool of air shaped like an inverted funnel, curving down from a thick, dark cumulonimbus cloud. If it touches ground, its lower portion darkens with soil and dust whipped up in the rotating column, and larger pieces of trees, fences, other shattered debris fall out around the funnel's bottom. Winds within and around the column whirl at speeds topping 300 miles an hour. Because of the centrifugal force effected by the rotating winds, air at the center of a twister thins and pressure drops to as little as half of normal atmospheric pressure, low enough to explode houses and other structures in the path.

As a tornado moves across land in a path often only a few miles long and as narrow as 100 yards, buildings, trees and other objects are destroyed, and pieces of wreckage drawn up into the vortex may be dropped to earth again miles away.

On the average, tornado paths are only a quarter of a mile wide and seldom more than 16 miles long, but there have been spectacular exceptions in which tornadoes have caused heavy destruction along paths more than a mile wide and 300 miles long. One tornado in 1917 traveled 293 miles across Illinois and Indiana, and lasted seven hours and twenty minutes. Its forward speed was 40 miles an hour, about average.

On April 3, 1974, 148 separate tornadoes caused 300 deaths, 5,000 injuries and many hundreds of millions of dollars damage between Decatur, Alabama, and Windsor, Ontario.

In 1931, a tornado in Minnesota carried an 83-ton railroad coach and its 117 passengers 80 feet through the air and dropped them in a ditch.

Tornadoes do their destructive work through the combination of winds and the partial vacuum formed in the center of the vortex. As a twister passes over a building, the winds twist and rip at the outside while the sudden great drop in pressure in the tornado eye leads to explosive overpressures within the building. Walls collapse or topple outward, windows explode, and debris is hurtled through the air in a dangerous barrage.

Tornadoes occur in all fifty states and in many parts of the world. But the area most conducive to their formation is the continental plains of North America. No season is free of them. Usually, the number of tornadoes is lowest in the United States during December and January, at peak in May. April, May and June are months of greatest frequency.

According to the National Oceanic and Atmospheric Administration (NOAA), in February, when tornado activity begins to increase, the center of greatest frequency lies over the central Gulf states. During March, the center moves eastward to the southeast Atlantic states, where tornado frequency peaks in April. During May, the center moves to the southern Plains states and in June northward to the northern Plains and Great Lakes area as far east as western New York State.

Why this drift? Tornadoes are generated with greatest frequency where masses of warm, moist air meet up with contrasting cool, dry masses. After May, when the Gulf states are largely exposed to warm air and relatively little cold-air intrusion occurs, tornado frequency drops. This is the case across the country after June. Winter cooling allows fewer encounters between warm and overriding cold systems, so tornado frequency reaches it lowest level by December.

The probability of a tornado striking any specific location is quite small. Even in areas most frequently subject to tornadoes, the probability of a strike at a specific location is about once in 250 years, and in the far western states the probability is close to zero.

But exceptions, of course, occur. Since 1892, Oklahoma City, for example, has been hit by tornadoes twenty-six times. In just one twenty-five-minute period on March 16, 1942, Baldwyn, Mississippi, was struck by two tornadoes. On May 30, 1879, fully one-third of Irving, Kansas, was left in ruins by two tornadoes that occurred forty-five minutes apart.

In the twenty-five-year period ending in 1973, the average number of tornadoes in the United States was 681 per year, with about half striking during April, May and June.

A tornado can occur at any hour of day or night, but because of the meteorological combinations that go into their making, they form most readily during the warmest hours of the day. About 82 percent occur between noon and midnight, and almost one-fourth (23 percent) of all tornado activity occurs between 4:00 P.M. and 6:00 P.M.

What causes tornadoes?

The storms form several thousand feet above the earth's surface, usually in conjunction with severe thunderstorms. These are not thunderstorms of the familiar local variety which occur when heating of the surface causes air to rise until cumulus clouds extend upward to great heights. Rather, tornadoes form along a squall line, a system of many thunderstorms. Such a squall line forms when a cold front pushes warm, moist surface air upward.

Typically, the warm air is marine tropical air coming from the Gulf of Mexico, flowing from the south, driven by circulation around a high over the east coast or the Atlantic. Above the warm air, there must be cold air moving at high speed from the southwest or west.

In order for a series of thunderstorms to develop in line within a narrow zone, a lifting mechanism is needed, and the daily warming of the earth's surface may provide it.

The warm air rises until, cooled by its own expansion, it loses buoyancy. Some of the air, however, will move up higher than the rest and, upon meeting colder, heavier air higher up, will become buoyant again in comparison and move up again. A number of such "chimneys" may form a squall line several hundred miles long.

It's along the southern edge of such a squall line that tornadoes may form. It's still not known exactly how. One theory has it that the rapid upward flow of warm air through the "chimneys" becomes a rotary flow and forms the tornado vortex.

Another theory is that something else besides the thermal effect is needed—a mechanical effect in which the rotating air currents are constrained by external forces, which lessen the radius of rotation. When this happens, the speed of rotation increases much in the way an ice skater can increase speed of rotation by drawing in his arms. Finally, the converging, accelerating rotary winds set up the tornado vortex.

Among the remaining mysteries about tornadoes is the relationship between tornado formation and thunderstorm electrification. It's possible that atmospheric electricity accelerates rotary winds to tornado velocities. On the other hand, it could be that the high-speed rotary winds generate electrical charges.

Although it's not possible to tell exactly when or where tornadoes will strike, it is possible to identify areas about 100 miles wide and 250 miles long in which weather conditions suggest a likelihood of tornado generation.

At NOAA's National Severe Storms Forecast Center in Kansas City, Missouri, weathermen constantly analyze atmospheric conditions over the forty-eight contiguous states, attempting to identify such areas.

Tornado watches—alerting messages—are issued by the center. They indicate the area and the period of time in which tornado probability is expected to be dangerously high. Watches are sent by Teletype to local offices of the National Weather Service and communicated to the public by radio and television stations in and around the endangered area. Law enforcement officers, emergency forces and voluntary storm reporters are also alerted by the watches.

A tornado watch is not the same as a tornado warning. A watch does not call for interrupting normal routines. It simply alerts you to be on the lookout for threatening weather.

A tornado warning is issued when a tornado has been sighted in the area or indicated by radar. Often, warnings are made possible through the work of SKYWARN volunteers who notify the nearest office of the National Weather Service or community warning center when a tornado is sighted. A warning includes the location of the tornado at the time of sighting, the area through which it may move and the time period during which the tornado will move through the warned area.

Here's one example of how the warning system works—and its value. A particularly violent tornado struck Lubbock, Texas, on May 11, 1970. That morning, forecasters at the Kansas City facility were concerned with atmospheric conditions over the high west Texas plains. Even before 10:00 A.M., there had been a forecast of weather-making vertical motion in the atmosphere. At 10:00 A.M., an amended forecast included isolated thunderstorms with large hail for the area. By early afternoon, the forecasters were advising area weather stations that the air mass over the area was unstable, that isolated thunderstorms would be severe, and local areas should be watched carefully for this possibility. At 8:40 the Kansas City weathermen were issuing a severe thunderstorm watch bulletin for the west Texas area.

Meanwhile, alerted by the morning notices from Kansas City, and by additional data received during the day from the National Meteorological Center near Washington, D.C., the Lubbock office of the National Weather Service was keeping a close watch, which intensified from midafternoon on during peak hours for tornado formation. By 6:00 P.M., towering cumuliform clouds in the area indicated increasing atmospheric instability. Shortly before 7:00 P.M., the local-use radar picked up a moderate thunderstorm 5 miles south of Lubbock. Continuous local radar monitoring in conjunction with reports from a larger radar system at the Amarillo weather station indicated rapid intensification of thunderstorms in the area.

By 7:50 P.M., a severe-thunderstorm warning was issued, and within minutes local radio stations began broadcasting it. At 8:08 a message was sent out not only reaffirming the severe-thunderstorm warning but also reporting egg-size hail south of the city. At 8:10 another message advised of grapefruit-size hail 5 miles south of the city. At 8:15, based on radar findings, a tornado warning bulletin was issued.

Within less than forty-five minutes radar picked up another tornado indication, and a report of a funnel cloud came in. Warnings were repeated during the next thirty minutes even as the tornado touched down in Lubbock at 9:35 and the weathermen had to take cover.

The tornado killed 27 people and injured 1,500 along its 8½-mile track. It wiped out 1,040 family units and damaged 8,876 more. Estimates of property damage exceeded $125 million.

The wonder of the Lubbock disaster, says NOAA, "was not that the city had been struck by a tornado, or that there had been casualties, but that there had been so few in a city of 150,000.

NOAA's work of timely warning, and the tornado-consciousness of Lubbock's citizens, living in the heart of tornado country, made the difference."

Tornado safety rules (these are advised by NOAA):

When a tornado watch is given, meaning tornadoes are expected to develop:
- Keep a battery-operated radio or television set nearby and listen for weather advisories—even if the sky is blue.

When a tornado warning is issued, meaning a tornado has actually been sighted or indicated by weather radar:
- Seek inside shelter—in a storm cellar or reinforced building—and stay away from windows. Curl up so your head and eyes are protected. Keep a battery-operated radio or TV nearby and listen for further advisories.
- In homes, the basement offers greatest safety. Seek shelter under sturdy furniture if possible. In homes without basements, take cover in the center part of the house, on the lowest floor, in a small room such as a closet or bathroom, or under sturdy furniture. Stay away from windows.
- In shopping centers, go to a designated shelter area—not to your parked car.
- In schools, follow advance plans and go to an interior hallway on the lowest floor. If the building is not of reinforced construction, go to a nearby one that is, or take cover outside on low, protected ground. Stay out of auditoriums, gymnasiums, and other structures with wide, free-span roofs.
- In tall buildings, go to an interior hallway, away from windows, or to the designated shelter area.
- In open country, move away from the tornado's path at a right angle. If there is no time to escape, lie flat in the nearest ditch or ravine.

FLOODS AND FLASH FLOODS

Each year the U.S. Weather Bureau turns out about 1.3 million general weather forecasts and 1.7 million aviation forecasts. It also issues about 75,000 river-stage and flood forecasts.

Floods, which are often disastrous, are virtually an inevitable part of life along rivers. On the average, each year some 75,000 Americans are driven from their homes, 90 are killed, and more than $250 million worth of property is damaged or destroyed by floods.

The cost of a flood to a community also includes possible danger to public health and intangible losses such as disruption of business and transportation.

Hundreds of times each year, tranquil rivers are transformed into destructive torrents, and no area of the United States is completely free of the threat.

Floods develop when soil and vegetation cannot absorb falling rain or melting snow and water runs off the land in such quantities that it cannot be carried in normal stream channels or retained in natural ponds and man-made reservoirs.

In the eastern part of the country, disastrous floods may occur when a low stalls in its eastward course and air in the warm part of the low is very moist and unstable and heavy, resulting in excessive rainfalls in advance of the warm front.

Generally, the worst floods occur in winter or early spring. Often, from December to March, heavy and continuous rains fall, and there may be melting snow as well. Summer rains can be heavy but they tend to be showery and scattered, and in summer the ground is better able to absorb and hold water. An exception is the torrential flooding rains that may develop with Atlantic hurricanes.

While the Atlantic and Gulf coasts can suffer considerable damage from floods, the greatest flood losses over a period of years tend to occur in the Mississippi Valley.

Flash floods—sudden local flooding—are typical of but not confined to the arid Southwest, in mountainous areas where the valleys and narrow canyons and the high slopes are relatively bare of trees and brush. A severe thunderstorm over the mountains may let loose a sudden deluge of water, which, tumbling into a dry gully, can transform it into a flooded maelstrom.

But flash floods can occur anywhere on small streams, especially near the headwaters of river basins, where water can rise quickly in a heavy rainstorm, and flooding may begin even before the rain stops falling.

Flash floods occur in mountainous areas when very heavy thunderstorm rains can suddenly turn a little brook into a torrent of water. In urban areas, where the flood plain has been converted for buildings, roads and parking lots, heavy rains can overload the drainage system, producing flash flooding that can wash away cars and cause damage to homes and factories at lower ground levels.

In 1972, according to NOAA records, torrential rains on the slopes of South Dakota's Black Hills caused catastrophic flash flooding along a two-block-wide, twelve-mile-long stretch of Rapid Creek, which flows through Rapid City, killing more than 200 people, and causing more than $100 million damage. Hardly two weeks later, the most widespread flooding in United States history was caused by hurricane Agnes in interaction with other atmospheric systems. As the hurricane moved up the Atlantic seaboard, its torrential rains caused flash floods—and most of the 122 fatalities—displaced entire communities and did an estimated $3 billion in damage.

The National Weather Service has a special river-and-rainfall reporting network and continually analyzes river-and-rainfall data to provide river forecasts and flood warnings in time to evacuate low-lying areas, move property to higher ground and take necessary emergency action.

The network includes river forecast centers that monitor meteorological and hydrological conditions affecting rivers and water supply. The centers provide water-level predictions for more than 2,500 points on the nation's rivers. Analyses are made of the past history of each stream and the relationships between precipitation, melting snow, soil conditions and stream flow. The analyses allow hydrologists to develop river-forecasting procedures for predicting the amount of water that will find its way into rivers and streams and the time that will take under various conditions of temperature, soil moisture and precipitation.

Also going into river forecasts is continuous information on present and expected atmospheric conditions supplied through special communication links by the National Meteorological Center in Washington, D.C.

Flood warnings can be issued hours to days in advance of the flood peak on major tributaries, and for main rivers, as far as several days or even weeks in advance. The larger the river, the more time between rainfall or snowmelt and rise in river height.

Flood warnings—forecasts of impending floods—are communicated to the public via radio and television and local emergency agencies. A warning indicates the affected river; whether the flooding will be minor, moderate or severe; and when and where flooding will begin.

The weather service has helped set up flash-flood warning systems in many communities. In these, a volunteer network of rainfall-and-river observing stations is established, and a local flood-warning representative is appointed to collect reports from the network and to issue official flash-flood warnings based upon these reports and their relationship to graphs from the weather service which show the local flooding that will occur under different conditions of soil moisture and rainfall. The representative can prepare a flood forecast and spread a warning within minutes.

Many communities within range of a National Weather Service radar system have an added means of getting advance warning when flash-flood-producing rains approach. For with modern improvements, radar can determine the location, speed and intensity of rainfall patterns.

Flood safety rules (these are urged by NOAA for individuals and communities):

Before the flood:
- Keep on hand materials like sandbags, plywood, plastic sheeting and lumber.
- Install check valves in building sewer traps to prevent floodwater from backing up in sewer drains.

- Arrange for auxiliary supplies for hospitals and other operations that are critically affected by power failure.
- Keep first-aid supplies at hand.
- Keep your automobile fueled; if electric power is cut off, filling stations may not be able to operate pumps for several days.
- Keep a stock of foods that require little cooking and no refrigeration; electric power may be interrupted.
- Keep a portable radio, emergency cooking equipment, lights and flashlights in working order.

When you receive a flood warning:
- Store drinking water in clean bathtubs and in various containers. Water service may be interrupted.
- If forced to leave your home, and time permits, move essential items to safe ground; fill tanks to keep them from floating away; grease immovable machinery.
- Move to a safe area before access is cut off by floodwater.

During the flood:
- Avoid areas subject to sudden flooding.
- Do not attempt to cross a flowing stream where water is above your knees.
- Do not attempt to drive over a flooded road—you can be stranded and trapped.

After the flood:
- Do not use fresh food that has come in contact with floodwaters.
- Test drinking water for potability; wells should be pumped out and the water tested before drinking.
- Seek any necessary medical care at nearest hospital. Food, clothing, shelter and first aid are available at Red Cross shelters.
- Do not come from outside to visit disaster area; your presence might hamper rescue and other emergency operations.
- Do not handle live electrical equipment in wet areas; electrical equipment should be checked and dried before returning to service.

- Use flashlights, not candles, lanterns or torches, to examine buildings; flammables may be inside.
- Report broken utility lines to appropriate authorities.

Flash Floods. Flash flood waves, moving at incredible speeds, can roll boulders, tear out trees, destroy buildings and bridges and scour out new channels. Killing walls of water can reach 10 to 20 feet.

When a flash-flood warning is issued for your area, or the moment you first realize that a flash flood is imminent, act quickly to save yourself. You may have only seconds.

- Get out of areas subject to flooding and avoid already flooded areas.
- Do not attempt to cross a flowing stream on foot where water is above your knees. If driving, know the depth of water in a dip before crossing. Remember that the road may not be intact under the water.
- If a vehicle stalls, abandon it immediately and seek higher ground—rapidly rising water may engulf vehicle and occupants and sweep them away.
- Be especially cautious at night when it is harder to recognize flood dangers.

And, as NOAA strongly emphasizes: *During any flood emergency, stay tuned to your radio or television station. Information from NOAA and civil emergency forces may save your life.*

THUNDERSTORMS

They have always been impressive, and even, for many people, anxiety provoking. The ancient Greeks worshipped Zeus as a god who not only wielded thunder and lightning, but visited earth on lightning bolts.

Certainly the amount of energy released in a thunderstorm is incredibly huge. Scientists have measured several hundred thousands of amperes of electricity in bolts of lightning—enough to service several hundred thousand homes.

123

What goes into making a thunderstorm?

In an ordinary rainstorm, very moist air is lifted only high enough for the moisture to be squeezed out and to fall as rain. In a thunderstorm, the moist air, because of atmospheric instability, is carried to a much greater height where it condenses into a thunderstorm cloud, a towering anvil-shaped cumulonimbus that can rise to higher than 30,000 feet.

Within the cloud, turbulent conditions can break up the raindrops so smaller droplets may rise to the top of the cloud while larger ones remain at lower levels, or the droplets sometimes freeze, producing small spicules of ice that rise to the top of the cloud.

As the result of such separation of the forms of moisture, there is also separation of electric charges. Positive electrical charges collect at the top of the thunderhead and negative charges accumulate below.

The negative charges and positive charges on the ground are mutually attracted. But nothing happens until somehow—and much yet remains to be learned about the intricate mechanisms of thunderstorms—a faintly luminous streamer arising from a region in the negatively charged base of the cloud forces an irregular channel only a few inches in diameter toward the earth. Within that channel, air is ionized, or made electrically conductive.

Then comes a fooler. When the channel reaches earth, a brilliant lightning flash develops, and lightning actually surges back along the streamer path, from positive to negative, from ground to cloud—not, as we often think, because the action is so swift, from cloud to earth.

The action is swift, indeed. A streamer moves toward earth at a rate of about 2,000 feet in about $\frac{1}{100}$ of a second, and the return lightning discharge moves at about 100 million feet a second.

Lightning strokes last only a few millionths of a second. Commonly several pulses follow each other, but so rapidly that to the eye they may seem like a single pulse.

Actually, only a relative few of the total lightning flashes involve the earth; most surge from one cloud to another in much the same way as those between earth and cloud base.

There's an old saying about lightning never striking twice in the same place—but it's not true. During a severe thunderstorm, it's not uncommon for the Empire State Building in New York, for example, to receive several hits.

Where does the thunder come from during a thunderstorm? On passing through the air, lightning produces great heat, with sudden expansion and contraction of the air, which causes vibrations and the shock, or sound, waves we hear as thunder.

Sound moves at a speed of about 1 mile in five seconds. And you can judge the approximate distance of lightning in miles by noting the time in seconds between the flash and the thunder, and dividing by five.

Scary as lightning may be, it has its beneficial effects. It's been estimated that something more than 15 million thunderstorms occur annually around the world, and that the lightning discharges may produce 100 million tons of fixed nitrogen compounds annually as they break down air and release nitrogen for deposition on soil and plants where it serves as a valuable fertilizer.

Lightning can kill, but is more likely to do so in rural than urban areas because in cities steel structures provide paths for lightning, and make better conductors than the human body.

Lightning, seeking the shortest route to earth, most often strikes the highest object in the neighborhood. Lightning rods protect structures by receiving the lightning and, through grounding, discharging it harmlessly to the earth.

In modern homes and large buildings, little chance exists of being hurt by lightning; so, too, in cars, buses or trains, which offer protection because of their metallic frames.

Thunderstorm safety rules (proposed by NOAA):
If you are caught outside, get inside:
- Go indoors, to a shelter if possible; the bigger the building the better.
- Get into an enclosed car, but not a convertible.
- If you must stay outside get away from isolated trees, metal pipes, wire clotheslines and fences, metal sports equipment such as golf clubs, tractors and all metal farm equipment.

- If you are at the beach, swimming or boating, get out of and away from the water.
- If you're caught in a wooded park go to a low spot where there are small trees and keep away from trees or small buildings that stand alone.

HAIL

Usually occurring in thunderstorms, hail consists of pellets or lumps of ice, usually about ¼ inch or less in diameter, but sometimes much bigger.

Hail occurs when large raindrops near the bottom of a thundercloud are caught in strong updrafts that carry them up to colder temperatures where they freeze. In the upper part of the cloud they gather snowflakes and ice crystals, grow larger and heavier, and begin to fall. Once again, even several more times, they may be caught in updrafts, be carried up, freeze again and collect more ice crystals.

Finally, they fall out of the cloud, bombarding the earth. Even the small ones can sometimes break windows, dent roofs of parked cars, destroy greenhouses and damage crops.

In one severe hailstorm over Amarillo, Texas, in June 1969, when stones with a diameter as large as 2 inches and a little more were seen, the damage was estimated at $15 million.

ICE AND FROST

Sometimes rain, in falling, passes through a layer of air near the ground with below-freezing temperature. In passing through the layer, the raindrops freeze and hit ground as small ice crystals or white pellets known as soft hail.

Such a fall can heavily coat and break down telephone wires and tree branches.

Dew, of course, forms when, at night, the ground cools to below air temperature and moisture in air near the ground condenses. When ground temperature is below freezing, white ice particles, know as hoar frost, or white frost, may be deposited instead of dew.

Snow originates in clouds at levels where temperatures are below freezing, allowing ice crystals to develop and mass together.

But that doesn't necessarily mean that snow will reach the ground. If, between the cloud level where it is formed and the earth, temperatures are high enough, the snow will melt as it falls and become rain.

We can get snow when temperatures near ground are a little above freezing, as high as 39 degrees Fahrenheit. At such temperatures, the flakes are likely to be large because partial melting takes place and smaller flakes stick together.

With temperatures well below freezing, snowflakes are usually small, dry and powdery.

Under some circumstances, when the temperature is too low for pure rain yet too high for pure snow, the result is the frozen rain we call sleet.

There are times when just a few hours of snowfall creates a thick covering blanket. At other times the same number of hours of snowfall may result in bare surfaces and wet ground. So it is obvious that snow is an unusual form of precipitation.

The snowflake is as varied as any pattern devised by nature. The flake under the microscope may be in the form of a six-rayed star, a hexagonal plate, a column, a box, a dendrite or a needle. The crystal may be symmetrical, plain, ornate or sculptured. The variety of flake form is one of the amazing characteristics of snow.

It was Johannes Kepler (1571–1630) who first drew attention to the characteristic six-sided symmetry of a snow crystal. And not long after that observation, René Descartes (1596–1650), the French mathematician, sketched a series of snow crystals and their unique structure.

Snow varies in its water content. One sometimes may hear a forecaster say "If it had been just a few degrees colder today, the one inch of rain that fell would have been ten inches of snow." That is usually the density of fresh fallen snow. In other words, if 21 inches of snow were melted down, one would expect to have the equivalent of 2.1 inches of water. And by the way, it may come as a surprise to you, but the snow depth which is measured by the National Weather Service at its stations, is a rough measurement. The observer seeks out a level piece of ground in the open and takes a series of depth measurements with a wooden ruler, which he sticks vertically through the snow to the ground. It's nothing more complicated than that . . . so you can compare the snowfall in your area with the official figures quite easily.

9

Toward Greater Accuracy

WHY ARE THERE still errors in the daily weather forecasts you see on TV and hear on radio? Why do forecasters differ among themselves?

If astronomers can predict to the minute eclipses of the sun and moon and the movements of planets, why is it not possible with similar accuracy to forecast, say, rain just one day in advance?

The answer of course lies in the complexity of weather, which is infinitely more difficult to get a handle on than celestial mechanics.

In weather, we're dealing with an atmosphere weighing several million billion tons, nearly all of it in motion, parts at the pace of a gentle breeze, parts at the furious speeds of a jet stream, and not only moving its own huge weight but also constantly picking up water by trillions of tons, hoisting it thousands of feet and dropping it in one form or another hundreds or thousands of miles away.

It's been said that if the only problem in weather forecasting were to predict the movement of high and low centers, and if these followed simple laws of interaction, then predicting weather might be no more difficult than predicting eclipses. But beyond predicting the movement of the centers at sea level, it's necessary to know their intensity there, plus their position and intensity at other levels, how pressure is distributed in between, the variations in temperature and humidity at various heights, wind velocity, cloud occurrence, precipitation.

All of these factors interact; the interactions are highly complex; we only partially understand the rules that determine how they interact. And while our understanding, though limited, is enough to allow some representation of those rules through mathematical

129

equations, the equations are so complex that to solve them fully would take computers of far greater speed and capacity than any now available.

Another limitation lies in the gaps in observations. Weather changes in any area are influenced not only by conditions in that area but in a large area surrounding it. Especially for longer term forecasts, but to some extent for predictions of tomorrow's weather, a more complete picture of present pressure, temperature, humidity, etc., throughout the troposphere and lower stratosphere over a large part of a hemisphere would be helpful. But observations are made by a limited number of stations, sometimes so far apart that important weather information which may lie between is missed.

Still, with all these difficulties and deficiencies, useful forecasts have been made for many years, with some improvement as to accuracy made recently, and more improvement likely in years to come.

It wasn't until the era of the telegraph that modern forecasting could even begin. Previously, there had been some very limited progress beyond what had for centuries been largely mythology. Charts were made, beginning about 1820—showing winds, pressures and temperatures—but they were based on the distant past. In fact, the first chart made in 1820 was actually based on observations made 37 years earlier. At least, from these came recognition that there were low pressures associated with bad weather and that these systems usually moved from west to east.

With the arrival of the telegraph, it became possible to collect information on a current basis and to begin to use the low-pressure concept in forecasting.

Not until 1870 did the United States have a national weather service and regular forecasting. At the start, the U.S. Weather Bureau was a part of the army and later of the Department of Agriculture. Now known as the National Weather Service, it is a branch of the National Oceanic and Atmospheric Administration in the Department of Commerce.

In those early years, here and abroad, forecasts steadily

improved as meteorologists began to acquire a kind of instinct for how patterns shifted on the charts from day to day.

Still it was not until World War I that Norwegian meteorologists made the discovery that the typical storms of both Europe and North America occur as the result of the collision of warm tropical and cold polar air.

Then came the radiosonde era, which started in the 1930s and experienced rapid technological development during World War II. As instrument-laden balloons made routine flights, sending back measurements of what was going on in air high above the ground, meteorologists were able to make greater sense of surface winds and conditions by connecting them with conditions aloft.

In the 1960s we entered a far more advanced era—and, undoubtedly, are only at its beginnings—an era not only of radiosondes but of satellites and the computer and of international cooperation.

Just how accurate are our forecasts today?

The answer varies depending upon the weather factors being predicted and the length of time covered by the prediction.

Generally, greatest accuracy is achieved in forecasting temperature and wind direction and speed. Predictions about cloudiness and precipitation measure up fairly well. But there can be substantial errors in just how much precipitation can be expected. And while forecasts for the next two or three days are often fairly accurate, the optimum prediction time covers only the next six to twenty-four hours. Beyond seventy-two hours, often the best that can be done is to offer general indications, although under some circumstances, with stable or very slow-moving conditions, specific predictions can hold up for as long as a week.

The National Weather Service, considering accuracy in terms only of predictions of precipitation, finds that these average 80 to 85 percent correct.

Can that record be improved? Undoubtedly. And there are now unprecedented efforts to make it possible.

One informed estimate is that, when it comes to a three-day forecast for the middle of the continent, perhaps as much as 45 percent of present error stems from computing procedures. Fast as modern computers are, they still have to take shortcuts in many calculations in order to get out forecasts quickly enough. But computer speed is not likely to remain an obstacle for long considering the pace of computer development.

Another source of error, according to that estimate, is lack of complete and accurate data on present weather around the world. That lack may account for about 20 percent of the total error. But if increasing the number of weather observation stations hugely to provide such data may not be a practical reality, there is considerable hope that increasingly effective satellites covering the whole globe can do the job.

Finally, about 35 percent of the total error stems from still insufficient understanding of all of the physical factors and processes that go into making weather, among many others, for example, the atmosphere's acquisition of moisture from the oceans, and what effects landscape texture has on winds, even on the transformation of clouds into precipitation. If increased and more effective use is to be made of computers in forecasting, meteorologists must understand such processes to the point of being able to give the computer specific instructions on the calculations required to take each process into account.

It is out of a conviction that such understanding can be obtained through more complete and more extensive observations of the global atmospheric structure that major international programs in meteorology have been undertaken recently. Many nations are cooperating in the Global Atmospheric Research Program (GARP) developed in the 1960s through the World Meteorological Organization. GARP's aim is to increase the range of accurate weather forecasts by as much as two weeks, develop better computer models of the atmosphere and possibly even to predict climatic change.

One concern of GARP is with a largely neglected area—the tropics. As J. T. Houghton, president of Britain's Royal Meteorological Society puts it: "If we want to understand how the atmosphere works, we have to be concerned with its boiler."

Until recently, meteorologists, frustrated by lack of adequate observations and mainly concerned with weather in the North Temperate Zone, generally have let the atmosphere's boiler—the tropics—take care of itself.

But now the tropics have become the scene of a massive research program. It began in 1969 with the Barbados Oceanographic and Atmospheric Experiment (BOMEX) in which United States and Canadian scientists used twenty-eight aircraft, a dozen ships and several satellites to study a 500-by-500-kilometer area of sea and air extending 5 kilometers up into the atmosphere and 500 meters beneath the sea.

Then in 1974, GARP set up its far more massive Atlantic Tropic Experiment (GATE) in which 5,000 investigators representing 72 countries used thirty-nine ships, thirteen aircraft, five satellites and some thousand land stations in an observation program covering some 50 million square kilometers of the tropical Atlantic between Africa and South America.

This is massive science and the payoff takes years. It took four years after completion of GATE before refined data became widely available to meteorologists who are even now trying to use the information to achieve new insights and increase knowledge.

In one important bit of work, J. S. Winston of the U.S. National Oceanic and Atmospheric Administration has been able to show how our recent severe winters in the Northern Hemisphere were foreshadowed by changes in tropical ocean-atmosphere energy flows and air circulation patterns.

Even such a huge enterprise as GATE, important as the data it produced will be, leaves room for more—which may well be provided by the most ambitious project yet: the $500 million First GARP Global Weather Experiment (FGGE) which dwarfed the $80 million GATE.

For a year, beginning in December 1978, 149 nations joined in a coordinated attempt to observe total global weather in detail, with even the People's Republic of China contributing two ships.

In the planning since 1973, FGGE for the whole year, including two 2-month "special observing periods" from February through

March and May through June, collected data from around the world, with special emphasis on the Southern Hemisphere and the tropics, on air and sea temperature, pressure and circulation. The observations were made by 5 geosynchronous satellites, 4 polar-orbiting satellites, 50 research vessels, 110 commercial and research aircraft, 300 constant-level balloons and 300 instrumented drifting buoys.

The program was so huge that as Verner Suomi of the University of Wisconsin, chairman of the United States' GARP committee has put it: "It's like giving the earth's atmosphere its first complete physical."

Actually, although the information-gathering program was initially expected to offer only long-term benefits, it was, even within six months of its start, transforming weather maps in Australia and pointing the way for improved forecasts in the United States. In addition, some aircraft have been able to save fuel and avoid storms because of information gathered in the experiment.

Both the immediate benefits to Australia and the potential ones for the United States derive from the success of the drifting buoys being used. Aviation benefits have come from weather information automatically taken by equipment installed in some wide-body jets and radioed back by satellite. The measurements, taken by the specially equipped planes on Southern Hemisphere routes where few weather observations are normally available, provide data that can be relayed to subsequent flights, allowing them to take advantage of favorable winds and avoid storms.

The drifting buoys, which measure the air pressure and temperatures of air and water and radio the data by satellite, were put into service in the Southern Hemisphere because there is so much water in that area and little land on which to locate weather-observing stations.

It was originally planned to store the data from the buoys and use it in later analyses of world weather. But the Australians made arrangements to get it quickly and use it in their forecasts. They had found that earlier information obtained from satellites was not wholly accurate; oceanic storm systems turned out to be much more violent and ocean currents south of Australia much stronger than was previously known.

A similar use of drifting buoys in the North Pacific could improve weather forecasts in the United States. The North Pacific has a major effect on weather in the United States, particularly in the West, generating the storms that bring rain to California, Oregon and Washington.

Scientists have speculated that immense pools of unusually warm and cold water in the North Pacific may have an impact on our climate. The pools, discovered in 1957, are 600 to 1,200 miles across and a few degrees warmer or cooler than normal for water in their location. Some scientists speculate that the pools, called anomalies, influence weather by steering both air and water currents. A buoy system could help scientists determine whether this is true and what the effect is.

Meanwhile there have been and continue to be other significant developments.

Satellites are being improved and being used for more and more purposes, some of them potentially very exciting.

The first weather satellite, Tiros I, went into orbit in 1960 and sent back pictures of the earth's cloud cover that amazed even meteorologists. Depressions and tropical storms stood out. Even now, observers are making new findings about weather processes by examining satellite pictures closely.

Some weather satellites fly on polar orbits—around the earth over the poles—so that with the earth's rotation the whole globe can be gradually scanned on successive orbits.

Others—the geosynchronous satellites—orbit around the equator at a height of about 22,000 miles, taking exactly twenty-four hours for a complete orbit, so that they keep in step with the earth's rotation and thus remain stationed over the same region. They can observe about one-fourth of the earth at once. The United States launched its first geosynchronous satellite in 1966. Others have been launched since.

The U.S. satellites are constantly being improved. They are now able to see the clouds and the earth's surface with much higher resolution than when the program started and now are able to look in different portions of the spectrum, such as the infrared, from a 23,000-mile distance with a temperature accuracy or resolution on the order of 1 to 2 degrees Celsius.

The satellites are being used in acquiring measurements of various forms of electromagnetic radiation that are converted into data on atmospheric winds, sea surface temperature, snow and ice cover on the earth's surface, the location and speed of storms, and even such things as the distribution of air temperatures from the earth's surface out to the stratosphere.

Especially important in the western part of the United States is the continuous monitoring by satellite of the snowpack by measuring the extent of the snow cover. The information helps provide a much better estimate of total water stored in the snowpack. Another role of the satellite is to assist in determining when the snowpack will start melting, providing some idea of the runoff. This helps improve water management by controlling reservoir levels and warns of flooding when there is rapid snow melting.

With infrared sensors, the geosynchronous, or geostationary, satellite can keep track of the frost line as it moves during winter nights. The information is used by the National Weather Service to provide improved warnings to citrus farmers in Florida.

Satellites are playing an increasingly important role in tracking hurricanes and tropical storms. Says David S. Johnson, director of the National Environmental Satellite Service of NOAA:

> We are able to estimate quite reliably the strength of these storms, which is very important in making forecasts of where they are going to go and how severe the damage might be. The geostationary satellites now provide essentially continuous monitoring. Another severe weather condition that is under continuous monitoring is tornado-breeding storms, or thunderstorms that can cause extreme damage and loss of life. . . .
>
> Before satellites, we had to rely upon the occasional ship or aircraft report to indicate that there was even a storm

around. Now, satellites detect the storms in their earliest, incipient state. They are under constant surveillance until they die and disappear. So they are never out of sight of at least one of our satellites. In the case of tornadoes, the satellite can give earlier warning of the development of the kind of thunderstorm that is most likely to cause damaging winds or a tornado, as well as more explicit information on such a storm's location and movement. There is considerable research going on using satellite, radar and other ground-based data to see if we can better understand what triggers a tornado. We are hopeful that these new observing techniques and theories now being developed will indeed lead to a breakthrough that would allow NOAA to increase the warning time of the tornado itself.

And looking ahead into the future of satellites, Johnson sees a number of developments coming, some that would have seemed almost inconceivable not many years ago.

I think we are going to find satellites being used much more extensively in other fields of the environment. Microwave sensors will allow us, perhaps, to measure the waves on the oceans around the world—their height and distribution, to be able to measure tides very accurately, to significantly increase our knowledge of the oceans.

We expect to be able to monitor important atmospheric components, such as ozone, that are of great concern to the health of the population and the effect of man in his environment.

I expect it will be possible to get very precise measurements related to climate and climate change, which may help answer some of the questions we have today about what is going to happen to the climate in the future, what man is doing to change his own climate and what kind of government action should be taken to avoid a disaster later on.

I can visualize that it won't be too long—maybe in five to ten years—before each of us will be able to carry on our wrist a radio receiver to get weather warnings all of the time. When we go home, or to our office, we will be able to punch a button on the TV set and see the latest satellite pictures with the clouds and the storms in motion. Superimposed on them probably will be a map of the radar echoes showing the areas where there is rainfall.

We will be able to say, "I guess I better get out to the parking lot and get into my car before that little storm hits me."

This is all feasible today. It is just a matter of money. But, with the revolution we have seen in the last 25 years, and no abatement in the development of lower and lower costs of solid state devices, I think that even money will not be a limitation in a matter of a very few years.

Actually, for several years, a research group at the University of Wisconsin has been working to develop methods for "nowcasting." That would be a new kind of weather service. It would go beyond present general weather forecasts, which are unable to predict individual clouds and showers, for example. Yet such small-scale weather can be important to individuals. And other small-scale weather—winter snowstorms, summer hailstorms—can be important to communities.

The aim of nowcasting is to produce from each new satellite picture, as it comes in and as it makes up part of a series tracking weather across the countryside, useful information about winds and weather for virtually instant dissemination to local weather stations and via television to the public. Under study are a whole series of ingenious map, picture and written information displays. Not least among them is an "instant replay" system, much like that used in televising sports events, that would show successions of satellite photos for the area so the viewer could see for himself the kind of weather heading his way.

And still other developments are in the making, some of them close to being here.

Rain radar. Basic research—investigations aimed at understanding nature and natural phenomena without any practical applications in mind—often pays off in all sciences. Here is one example of that in meteorology.

Several years ago, at Ohio State University, Dr. Thomas A. Seliga, an associate professor of electrical engineering, and a colleague, Dr. V. N. Bringi, carried out fundamental studies on raindrops. Among other things, they were interested in raindrop shape.

Were raindrops really spherical? No; it turned out that, in fact, the drops are more tear shaped.

Of no consequence? On the contrary, four years later, the two engineers had developed a new radar technique that can measure the size and number of falling raindrops.

The technique is based on use of radars beamed along two planes, vertical and horizontal, so that the difference in radar reflection gives a measure of drop size and number. And the radar reflection is different in the different planes exactly because the drops are somewhat tear shaped rather than spherical.

There now have been successful experiments with the technique over a two-year period. And it promises to be valuable in providing more accurate estimates of the rainfall in storms. That would be of great help in, among other things, flood prediction and the solving of many water-related problems.

Snowpack probing. Thanks to developments in microwave technology, we may soon have at work automatic sensor stations that can monitor mountain snowpacks and warn of possible avalanches, and radarlike instruments that can predict the spring and summer runoff of a snowpack.

For three winters in succession, scientists from the National Bureau of Standards of the Commerce Department have moved out of their warm laboratories in Boulder, Colorado, headed for the Rocky Mountains. Wearing heavy parkas, thermal underwear, mittens and snowshoes, they have hauled microwave analysis and measuring equipment by toboggan into mountain passes and other "backcountry" areas to make microwave measurements of the snowpack, to determine its composition and moisture content.

The potential payoff is great. Given accurate measurements of moisture content of snow, there can be more accurate predictions of spring and summer runoffs—something vitally important in the western United States where about half of all irrigation water comes from the mountain snowpack runoff.

Analysis of snowpack composition can also be a tip-off to avalanche conditions. Given accurate analyses, it becomes possible to establish a network of microwave sensors connected to a central data-receiving office to warn of avalanche conditions.

There has already been considerable success in measuring the water equivalency of alpine (Rocky Mountain) snow, with the correlations between actual water content and microwave measurements coming within 5 percent.

Weather-mapping ocean eddies. Recently, in a joint effort of American and Soviet scientists called POLYMODE, a small fleet of research ships and hundreds of unmanned submerged instruments were used to make "weather maps" showing the development and evolution of eddies and currents in a 300-mile square of the Atlantic Ocean centered 300 miles southwest of Bermuda.

The POLYMODE weather maps, which look very much like surface weather maps, show highs, lows and lines of equal temperature. The "highs" and "lows" on the eddy maps correspond to temperature instead of barometric pressure as they do on a surface weather map. A daily series of maps shows the movement of the eddies.

Eddies in the ocean are huge whirlpools between 50 and 200 kilometers in diameter. Often called "mesoscale eddies," they represent one of the major ways the ocean stores and transfers energy, momentum and heat.

Why the interest? Says Robert H. Heinmiller, a research associate in meteorology at the Massachusetts Institute of Technology and U.S. manager of the POLYMODE Program: "We are certain that there is a great amount of interaction between the sea and the atmosphere. And we hope one day to be able to use that relationship to help in predicting the weather. More information about eddies would also help the fishing industry and the Navy."

Eddies differ in size, intensity and structure. They appear to be random and constantly changing. In addition to their roughly circular motion, all in the section of ocean studied move slowly to the west. Before reaching the east coast of the Americas, they presumably dissipate.

Superficially a picture of the eddies would look like the swirls that appear in a coffee cup when cream is poured in—except the physical mechanisms are different, and the cup is the size of an ocean basin and the eddies are miles across. Although they move slowly, at a quarter mile an hour, they are massive, with some 60 trillion tons of water in a 100-mile wide eddy.

Until relatively recently, physical oceanographers assumed that the deep parts of the mid-ocean were basically just huge, slowly moving pools. But as technology improved and measurements could be taken, the regions were found to be anything but tranquil. The purpose of POLYMODE is to discover as much as possible about the eddies—why they are there, where they go and what effects they have.

Prediction models. It has long been thought among those who develop and run weather prediction models that what is happening in South America or Australia doesn't matter when it comes to forecasting ultra-long atmospheric waves—as opposed to the short-wave highs and lows we experience daily—in the Northern Hemisphere for the next week.

And because of that belief, meteorologists who make short-term predictions of large-scale circulation in the atmosphere of the Northern Hemisphere have relied on models that include no data at all from the Southern Hemisphere.

But now that view has been shaken.

At a recent American Geophysical Union meeting in Washington, D.C., Richard C. J. Somerville of the National Center for Atmospheric Research in Boulder, Colorado, reported what he called "convincing evidence" that the biggest waves are affected "dramatically and drastically" within a day or two by the tropics.

These large-scale waves—only three to six occur around the globe—are crucial to long-range forecasting. In such waves, "information" is stored and ultimately passed on to the short atmospheric waves—the highs and lows that determine if it rains or the sun shines at any given spot on earth.

And, as Somerville notes, anything that can make forecasts of large-scale circulation better "is money in the bank" for better daily weather forecasting.

Somerville set about comparing the long-range forecasting accuracy of a number of different prediction models. Using the actual conditions at Boulder for verification, he found that one model—which actually considered the whole globe and made use of global data—produced remarkably accurate five-day forecasts. On the other hand, a model considering only the Northern Hemisphere was good for only one to two days while another that included some data for the Southern Hemisphere turned out accurate forecasts for three days.

The results of Somerville's study come at a particularly propitious time in view of the masses of data from the tropics collected by the GARP Global Weather Experiment. Very soon after Somerville announced his research, the National Meteorological Center of NOAA, which has been using a prediction model limited to the Northern Hemisphere, began to plan for quickly testing a global model.

Sun and weather. Ever since Benjamin Franklin reportedly nearly electrocuted himself in the process of establishing that lightning is electricity, it has, of course, been known that electricity is involved in weather. And not only has it long been suspected that the sun is also involved, but there has been ever-increasing statistical evidence correlating solar cycles and earth weather cycles. But until now scientists have been hard put to propose a physical connection between the two kinds of cycles for lack of any mechanism that might explain how they were interconnected.

Heat might be a factor. But thermal power reaching the earth from the sun varies less than 1 percent, hardly enough to help explain the phenomena.

Moreover, the solar cycles to which earth weather cycles appear to be related clearly involve such electrical and magnetic phenomena as sunspots and solar flares and even the earth's movement through sections of the sun's magnetic field.

Now Ralph Markson of Massachusetts Institute of Technology has proposed a theory that the relationship is indeed electrical and involves thunderstorms.

The theory is somewhat revolutionary, a bit complex, but not beyond understanding by lay people.

It comes down to this: According to Markson, there is a kind of global electrical circuit and electrical storms act as the generator in the circuit. The earth and the atmosphere form a kind of capacitor for holding and storing electric charges. The "outer conductor" is an equipotential surface at about an altitude of 60 kilometers and the inner conductor is the surface of the earth. Between the two conductors, the atmosphere acts as a dielectric, or nonconductor, but a somewhat leaky one.

When current is generated by storms, it flows from the cloud tops to the ionosphere. And the return part of the circuit, from ionosphere to earth in fair weather areas, involves diffusion of ions—charged particles—through the air, with positive ions diffusing downward and negative upward.

In this global circuit, according to Markson, most of the resistance is in the generator. A million or more ohms is concentrated in the generator, and this represents 90 percent of the resistance in a circuit involving 1,500 thunderstorms generating an ampere of current each.

According to Markson's theory, this means that much of the resistance can be affected by changes in atmospheric conductivity. And these changes come about through changes in the ionizing radiation reaching earth from space.

Solar phenomena can change the earth's environment in one way or another to alter the amounts of ionizing radiation reaching the upper atmosphere. When there is an increase in ionization, the result is more thunderstorms. And thunderstorms release tremendous energy, about 100 million kilowatt-hours each on average. An increase in this energy could affect the global atmospheric circulation and weather and climate cycles.

It remains, of course, for Markson's hypothesis to be tested, as any hypothesis must be. And Markson suggests a good reason for the testing to be done. If the hypothesis turns out to be correct, we would have a possible way, by nuclear explosions or other means, to induce changes in upper atmospheric ionization that could modify the weather.

And changing the weather is hardly beyond the ability of man. Inadvertently, we have long brought about certain changes and we keep experimenting with deliberate efforts. We'll take a look at both in the next chapter.

10

How We Change the Weather

THE FAMILIAR old saw has it that everybody talks about the
weather but nobody can do much about it. Not quite true.

Man has influence on the weather—so far, perhaps, more
unwittingly than deliberately. Yet deliberate efforts to modify
weather have been underway in this country since shortly after the
Second World War. Some of them—cloud seeding for rainmaking,
for example—have had considerable publicity and a lot of ups and
downs. We'll come to those. But let's begin with the unplanned
human influences on weather.

UNINTENTIONAL WEATHER MODIFICATION

Urban-related weather anomalies. During the summers of 1971
and 1975, a large multi-agency research project was carried out in
the Saint Louis region. Called METROMEX (for Metropolitan
Meteorological Experiment), its objective was to determine the
effects of a large urban-industrial complex on the weather around
and downwind of it. And the project found a substantial number of
effects: increased cloudiness (up 10 percent), increased summer
rainfall (up 30 percent) and increased frequency of severe thunderstorms (up as much as 100 percent). The increases were found
to occur over and just east (downwind) of Saint Louis in a fan-
shaped area of 4,000 square kilometers.

There have also been effects of the effects.

The study found that the summer rainfall increases have led to
increases in the corn and soybean yields in the area east of Saint
Louis. Even after taking into account the extra hail and wind
damage caused by the additional storms, the net agricultural
benefit to the area affected by the altered precipitation has been a

145

2- to 5-percent increase in grain crop yields, worth about $1.6 million. As a result, agricultural land values increased. On the other hand, the more intense urban rains have led to 15 percent more runoff, 50 to 100 percent more local flooding and as much as 400 percent more stream and ground-water pollution.

A major reason for the difference between city and rural climate is that city air is markedly warmer. It has been estimated, for example, that New York City generates seven times more heat than it receives from the sun.

Just to begin with, cities—with their asphalt, brick, concrete and stone—can store more heat and store it more rapidly than rural areas with their grass, leaves and bare soil. The very structure of cities leads to greater absorption of sunlight. Rural areas consist largely of horizontal surfaces that immediately reflect much sunlight back to the sky, although some is absorbed and turned into heat. In cities, however, with their many vertical structures, sunlight can be reflected from vertical wall to horizontal and back to vertical wall again, with more energy absorbed in the process.

On top of this, of course, comes the enormous generation of heat in urban concentrations—heat escaping from very imperfectly insulated homes, stores, factories, power stations, and from the radiators and exhausts of vehicles. Nor does heat generation stop in summer when air conditioners remove heat from buildings and dump it outside.

Moreover, rural areas have much better cooling processes than urban. In the country, much of the rainfall is absorbed into the ground and then returned to the air by evaporation and by transpiration through plant leaves, both of which absorb heat. In urban areas, however, rainfall moves from roofs and streets into sewer systems, and little is available for evaporative cooling. As you might expect, relative humidity in cities is higher than in the country because of the reduced evaporation and the higher temperatures.

Rain in the city is greater than in the country—and so, too, the

146

cloudiness—because of the updrafts over the warmer urban area.

Fog is greater in the city because of dust and smoke. On calm days, when the dust and smoke are not blown downwind but hang over the city, the particles may become nuclei, water droplets may form on them, producing fog and weighing the nuclei down, holding them in the area until a good wind or heavy rain comes along, meanwhile allowing a haze to build up.

So human heat sources affect local climate and they may affect regional climate as well. And there has been some suggestion that urban concentration in the eastern United States might be a factor in attracting Atlantic storms to the coast.

Man-made deserts. Forest fires, insects, and plant diseases have taken their toll in destroying forests and natural vegetation, sometimes producing dust bowls. But over many thousands of years, farmers, clearing vast acreages for cultivation, have produced some of the same results.

There is a theory that destruction of vegetation may act to encourage the sinking of air and that revegetation of at least some of the desert areas may lead to increases in local rainfall.

Dams, which create artificial lakes, help to reduce the flooding caused by heavy rains. They may also reduce vegetation downstream—or, on the other hand, may help to increase it through irrigation projects.

According to theory, dams should increase rainfall over the nearby area.

Aircraft. The condensation trails—artificial clouds—produced by high-flying aircraft are familiar. There has been concern over the release by jet engines higher up—into the thin air of the stratosphere—of water vapor, carbon dioxide and nitrogen compounds. A particular concern is that the reactive substances released, especially from supersonic aircraft, may destroy in part the ozone layer in the stratosphere which keeps it warm and screens out potentially lethal ultraviolet sun rays so they don't reach the earth. Intensive research into this is underway.

147

The idea of modifying the weather is hardly new. In ancient times, attempts were made to influence it by prayers and even sacrifices to the gods.

In modern times there have been many ambitious if not practical proposals. There have been suggestions that the general circulation of the atmosphere could be modified by introducing clouds of soot or ice crystals at particular heights and latitudes in order to change the amount of radiation arriving on earth. There have also been proposals to put a carbon black coating on snow and ice fields at high latitudes and to eliminate the ice from the Arctic Ocean by pumping warmer Atlantic Ocean water into it.

Understandably, these have remained entirely speculative proposals both because of the huge investment of money, materials and energy they would require—and, no less, because the full consequences cannot be predicted.

Rain. The only substantial efforts to modify the weather have been aimed at promoting rain, reducing hail or moderating storms by cloud seeding. Rain promotion has had among its purposes increasing crop yields in arid or semiarid regions or in areas undergoing drought, augmenting water supply for domestic or commercial purposes, or increasing water flow for hydroelectric power generation.

You'll recall that clouds develop when moist air rises and reaches a height at which it becomes too cold to keep its water vapor. But the water droplets need some kind of nuclei, such as ice crystals, on which they can condense if rain is to fall.

In 1946, two scientists working for General Electric in Schenectady, New York, pioneered cloud seeding. Vincent Schaefer found that pellets of dry ice (solid carbon dioxide) dropped through a supercooled cloud could promote ice formation—and Bernard Vonnegut found that silver iodide could do the same.

Dry ice can be sown into a cloud from airplanes. Silver iodide can be released into a cloud as smoke, sometimes from an airplane, sometimes from the ground if necessary winds and updrafts are present.

Cloud seeding was launched with high optimism. Even now, however, there is no definitive answer as to how effective it is. Few meteorologists doubt that the implanting of chemical nuclei—and sometimes the seeding may be done with other materials such as salt crystals or even plain dirt—can actually lead to precipitation. But whether seeded clouds produce more moisture than they would naturally is extremely controversial. The problem lies with the near impossibility of carrying out well-controlled experiments on seeding.

In recent years, too, it has become clear that cloud seeding can produce complicated effects. For example, rainmaking may fail both because of too little or too much seeding. Rainfall can actually be reduced by seeding under some conditions. Also, some researchers now believe that both increases and decreases in rain may extend far beyond the area directly seeded. Here again reliable evaluation of these effects is hampered by the limited ability of experimenters to predict how a cloud would have behaved if it had not been seeded.

Many meteorologists believe that they can enhance rain- and snowfall by picking their clouds carefully. Scientists who seeded clouds in Colorado from thirty-five ground generators for five years have concluded that they may have achieved seasonal snowfall increases of 10 percent and possibly more. On the other hand, rainfall actually decreased during earlier seeding projects in Missouri and Arizona.

In 1978, a report to the congressionally mandated Weather Modification Advisory Board from its Statistical Task Force noted that a strict evaluation of seven rainmaking experiments over a period of five years left only one that the task force found statistically convincing. This experiment, the second of two successful Israeli experiments, appeared to achieve a 15 percent increase in rainfall. The other experiments yielded results that fell short of statistical significance or they were so questionably designed that their interpretation was confused.

Rainmakers often face another problem: an empty sky. "When you need rain the worst," laments one, "there aren't any clouds to seed."

Another issue involves the downwind effects of cloud seeding. In 1977, Idaho's attorney general threatened to sue the state of Washington, arguing that by seeding passing clouds it was in effect stealing Idaho's future rainwater. The attorney general was dissuaded by the governor from pressing the suit, but the question remains.

Some experts say that there is only so much moisture in the atmosphere and if you take it out in one place, it is gone. Others disagree, pointing out that only a small percentage of the moisture passing over a state or region falls as precipitation there.

The crucial point in any lawsuit would be evidence that weather modification in one area causes drought—or floods—in another region. So far, no such case has been made. It could be, of course, that as rainmakers learn more about their craft, they might just possibly supply scientific evidence against themselves.

It has been suggested by some authorities that an internationally planned experiment that would ensure the cooperation of respected meteorologists from around the world might offer an answer to rainmaking—Does It Increase Rainfall or Not? The World Meteorological Organization is conducting a feasibility study on a location in Spain. The results are not in as this book appears in print, but perhaps such a joint international venture would provide some conclusive evidence.

Fog. Seeding has been used with some success in dissipating fog, especially supercooled fog. At many airports where fogs in winter are frequent, small planes take off when fog that might close down operations is expected.

At a distance upwind, the planes release dry-ice pellets to create a hole in the fog. If wind does not carry more fog in, the planes land; otherwise they carry out more seeding.

In some cases, the clearing of fog is done from the ground by the release of liquid propane through orifices. As the propane evaporates after release, it cools to produce ice crystals.

For many airports, dissipating warm fog is more important because more hours of decreased visibility occur at temperatures above freezing. There has been some limited success in clearing warm fog by seeding and by employing helicopters to fly across the top of the fog to introduce warmer, drier air in the downwash into the cloud.

Hail. A key characteristic of hail in the United States is its enormous variability. Most areas experience only two or three hailstorms a year, and only 5 to 10 percent of these storms may produce seriously damaging hail.

During the warm months of the year—April through October—crop-damaging hail falls somewhere in the eastern two-thirds of the country almost every day. On twenty days in an average year, crop losses from hail exceed $1 million.

Although most damage from hail is to crops—and averages $773 million annually—in addition, property damage is done that runs to about $75 million yearly.

Half of all hail losses occur in the Great Plains—from Texas to North Dakota—where hailstorms are intense. Insurance is not a complete solution. In some areas where hail losses are high, many farmers can't afford insurance, and the insurance industry finds it difficult to price coverage at a profitable level.

A basic concept of hail suppression is what scientists call "beneficial competition." The idea is to introduce enough artificially induced hailstone embryos—the basic cores on which hailstones grow—so that the competition for the water supply among all embryos may be great enough to prevent any from growing to large size.

The artificial embryos are produced by seeding the cloud with an ice-nucleating agent such as silver iodide which can either freeze supercooled cloud droplets and convert them to ice crystals or grow ice crystals from the vapor. The crystals then grow into harmless snow pellets.

Actually, attempts were made to prevent hail long before the introduction of seeding. In France, Italy and Switzerland, gunfire and rockets were used and some success claimed, although no good explanation has ever been offered as to why these methods should work.

And as for the success of seeding methods, scientific knowledge is still incomplete. Tests of seeding to reduce hail in several countries have led to variable results. A program in France from 1959 to 1966 produced a decrease of 22.6 percent in annual loss due to hail damage compared with the previous fifteen years. There have been claims from the Soviet Union of remarkable success—hail reductions ranging between 70 and 90 percent. But there have been admissions in the Soviet literature of difficulty in suppressing hail in very severe storms, and scientists have questioned some of the reports.

Here in the United States, several hail-suppression programs have shown promise. In one, over a period of four seasons in two west Texas counties, there appears to have been a 48 percent reduction in crop loss. Over a period of fifteen seasons in two southwestern North Dakota counties, the reduction was 31 percent for crop hail losses, on which insurance rates are based, compared to nearby unseeded counties.

Given intensive enough study of hailstorm structure and the seeding of the storms, improved methods of hail control are expected.

Lightning. Lightning is a major cause of forest fires, particularly in the western United States where it ignites about 10,000 a year. The possibility that the destruction could be reduced by cloud seeding has, of course, occurred to the U.S. Forest Service. In studies by the service in the Northern Rockies, 66 percent fewer cloud-to-ground discharges, 50 percent fewer intracloud discharges, and 54 percent less total storm lightning have been noted during seeded storms than during unseeded.

Of possibly even more significance, there were also decreases in the number of cloud-to-ground flashes over a given period, as well as decreases in the average duration of discrete discharges and the average duration of continuing current in discharges. Discharges with long-continuing current are responsible for most forest fires.

Some scientists believe that the modification of the nature of the discharge may be more important than any change in the total lightning produced by a storm.

Hurricanes. Attempts to reduce the severity of hurricanes are based on the idea of seeding the outside edge of the eye wall with silver iodide in order to broaden the area over which heat is released, increasing the distance over which pressure decreases, thus weakening the winds and the whole storm.

After very preliminary experiments with Hurricanes Esther in 1961 and Beulah in 1963, a major study was made with Hurricane Debbie in August 1969. Aircraft seeded the storm on August 18, and the maximum wind at 12,000 feet did drop from 98 to 68 knots. On the following day, the storm recovered and on the third day, August 20, seeding was carried out again, with a wind reduction from 99 to 84 knots.

It's possible that with further study, taming of hurricanes might be achieved. But there are critics. Some believe that the seeming success with Debbie wasn't a success at all. They consider that the storm could easily have diminished naturally as it ran over cool patches of sea water. Skeptically, they ask: Considering the huge natural release of energy in a hurricane—something on the order of an H-bomb being exploded once a minute—how can it be expected to respond to, or even notice, a few pounds of silver iodide?

Others worry that hurricanes may be an essential part of the worldwide weather machine, and that interfering with them is not something to be undertaken lightly. Control Atlantic hurricanes, for example, and it is conceivable, they say, that Europe's rainfall might be reduced.

Japan worries about such interference. A quarter of Japan's rainfall is estimated to come from typhoons. And although the Japanese are exposed, on the average, to four typhoons a year, with the average damage from each one running to $100 million, the ill wind does a lot of good for farmers. So Japan is prepared to put up with typhoons while it uses an extensive early-warning system.

* * *

Certainly, then, the potential for weather modification exists but much research remains to be done before man can truly control the weather.

In 1978, in its report to the Secretary of Commerce on "The Management of Weather Resources," the congressionally mandated Weather Modification Advisory Board expressed optimism. It declared that, with more money and hard work, significant modification of many kinds of weather seems to be probable in the next two decades. It estimated that snow in the mountains and rain in the High Plains and Midwest could be increased by 10 to 30 percent by the late 1980s, and that hail reduction of up to 60 percent in some kinds of storms could be realized by the 1990s.

Some meteorologists visualize a kind of ultimate ideal weather service, of which weather control would be part. They see data from satellites, balloons and surface weather stations being fed to computers automatically, with the computers processing them to predict conditions to be expected if no modification treatment were used. And the computer, all automatically, would then go on to weigh social, economic and other benefits, determine the treatment required to produce optimal weather conditions for the most people, then set in motion the processes needed for the modification.

But clearly, right now, we are a long way from such a Utopian state.

11

Are Major Weather/Climate Changes Coming?

• In April 1974, the most devastating outbreak of tornadoes ever recorded, 148 of them, killed more than 300 people and caused half-a-billion dollars' worth of damage in thirteen states.

• In England, since 1950, the growing season has shortened by about two weeks, with an estimated loss to farmers of 100,000 tons a year in grain production.

• Since 1950, the average temperature around the equator has risen by a fraction of a degree—enough so that in some areas it can mean drought.

• In 1976, western Europe recorded one of the hottest, driest summers in a century. Farmers in England, northern France, Belgium, northern Italy and West Germany went months without rain, their fields drying out and crops shriveling. In 1976, drought hit the United States, too, with drought-caused crop losses in Wisconsin alone estimated at $400 million.

Do such events mean that something is happening to world climate? Climatologists don't really know. Some think the earth may be in a cooling trend, that after three-quarters of a century of remarkably mild conditions, we may be moving toward a kind of "little ice age," with generally cold and damp weather such as prevailed between about the beginning of the seventeenth century and the middle of the nineteenth century.

If some of the most pessimistic are correct, and we are headed for a climatic change as profound as they fear, there might be catastrophic famines. Several years ago a report by the National Academy of Sciences noted: "A major climatic change would force economic and social adjustments on a world-wide scale because the global patterns of food production and population that have

155

evolved are implicitly dependent on the climate of the present century."

Several studies have shown a half-degree drop in average ground temperatures in the Northern Hemisphere between 1945 and 1968 (one study, by a German climatologist, indicating that the atmosphere in the Northern Hemisphere was a full degree centigrade cooler in 1972 than in 1949); a sudden large increase in Northern Hemisphere snow cover in the winter of 1971/72; a drop in the amount of sunshine reaching the ground in continental United States of 1.3 percent between 1964 and 1972.

If, indeed, the cooling theory is right, the prospects would not be pleasant. With an average global temperature drop of only one degree, growing seasons in temperate zones could be shortened by a week or more, reducing food supplies. And energy sources might be further strained by increased heating requirements.

But other experts believe that if there is any cooling going on, it is being offset over the long term by an increase of carbon dioxide in the atmosphere from the burning of fossil fuels which is having a "greenhouse effect," warming the atmosphere by preventing some of the heat radiated by the surface of the earth from escaping out into space.

Those who believe there is a long-term warming effect point to recent uncharacteristically mild winters in such areas as Scandinavia and New England and a retreat of glaciers in the Alps.

Any such warming, even if it were relatively modest, also could have detrimental effects. The melting of polar ice could raise ocean levels, not only introducing danger of extensive flooding but also of marked changes in global circulation of air and in patterns of rainfall, with possibly a significant decline in productivity in many important agricultural areas.

Based on the available information, we seem to be living in a particularly benign era—in the, apparently, warmest century of the last millennium, during the warmest 10,000-year period out of the last million years.

From what is known, it appears that major ice ages, during which there may be temperature drops of as much as 8 degrees

Celsius, come about every 100,000 years, and since our current "Holocene" interglacial (warm) period began only about 10,000 years ago, no such major change would seem likely.

Within the great epochs, there appear to be other cycles, one of about 20,000 years and the other of about 2,500 years duration, during which there are temperature changes on the order of 3 and 2 degrees Celsius, respectively.

The evidence suggests that the shorter of those two cycles may have reached its most recent nadir in the "little ice age" that extended several hundred years to the mid-nineteenth century, during which paintings showed glaciers extending down whole mountainsides in the French Alps, mountainsides now covered with trees.

So it is possible that we may be in an upward swing of the shorter cycle, approaching the end of a warm period in the longer one. And, as the National Academy of Sciences report of several years ago noted, "The question naturally arises as to whether we are indeed on the brink of a [10,000 year] period of colder climate."

In its report, the academy, concerned with the need for a sweeping new research program, went on to concede that what causes the onset of major and minor ice ages remains a mystery. "Our knowledge of the mechanisms of climatic change is at least as fragmentary as our data. Not only are the basic scientific questions largely unanswered, but in many cases we do not yet know enough to pose the key questions."

If climatologists agree on anything it is that whatever may be the actual long-term trend, we seem to be entering a period of increasing climate variability that could make prediction—and planning—even more difficult.

"I do not see glacial melts or an ice age," says Dr. Jerome Namias of the Scripps Institute of Oceanography, La Jolla, California. "What I see is fluctuation."

Climatology is certainly a very young science and one that faces remarkable complexities.

You'll recall that the atmosphere is basically a heat-exchange vehicle. Heated air over tropical areas rises and moves toward the poles. As the air cools and descends, high pressure systems develop, with winds circulating clockwise in the Northern Hemisphere because of the earth's spin. And out of the interaction of such systems develop the major movements of wind and weather.

It has been said, quite aptly, that the factors that can modify this basic pattern may be the most complex and interrelated of any natural phenomena scientists have ever tried to study.

And to make matters even more complex, there are man-made activities that can enter into the picture.

One review in *Science News* a few years ago noted some of the many factors entering into climate. For one thing, the air's chemical composition determines how much sunlight is absorbed, with carbon dioxide increasing absorption and small particles capable of either increasing or decreasing absorption, depending on their nature. The distribution of clouds can affect the amount of light reaching the ground. Growth of polar ice caps can change the proportion of light reflected back into space. Extensive volcanic eruptions appear to be involved in climate. Sunspots, the wandering of the earth's poles, continental drift, and changes of the earth's position relative to the sun and other planets can all have an effect.

ORBITAL VARIATIONS

For a long time, there has been a suspicion among many scientists that a major factor in variations in the earth's climate lies in the subtle, regular variations in the earth's orbit around the sun. Recently the influence of those variations has gained increasing scientific acceptance.

The first suggestion that orbital variations might affect climate appears to have been made as far back as 1830 by John Herschel, an astronomer. The idea wasn't forgotten and came up several times over the years. Most recently, it was revived by a Serbian geophysicist, Milutin Milankovitch, who in 1941 published detailed calculations of orbital changes and offered a detailed theory as to their effects. But the theory had to wait for validation by substantial physical evidence—detailed chronology of global climatic reversals.

Recently, considerable evidence has been found by James Hays of Columbia University, John Imbrie of Brown University, and N. J. Shackleton of Cambridge University, England, all participants in an international project known as CLIMAP (Climate: Long-range Investigation Mapping and Prediction).

According to the theory, regular, easily predicted changes in the orientation of the earth's axis of rotation and the shape of its orbit affect sunlight distribution over the earth, and this varying distribution controls the timing of glacial epochs.

In more detail, the theory holds that three overlapping cycles of variations in orbit combine to change climate.

In the shortest cycle—some 23,000 years—the earth advances in its elliptical orbit so that it approaches closest to the sun at different times of the year. Currently, the earth and sun are closest in January. In another 10,000 years they can be expected to be closest in July.

When, as now, there is a longer distance between earth and sun in summer, temperatures are cooler, less snow melts and the polar ice caps tend to grow—all of which should slowly take us into a new ice age.

In the second cycle, of 41,000 years duration, there is a tilting of the orbit and the earth's axis sometimes is more nearly perpendicular to the direction of the sun than at other times. It appears that it was some 9,000 years ago when the axis tilt away from the perpendicular was greatest. The present trend toward minimum tilt also favors a new ice age, since as the axis moves toward that minimum tilt—about 22 degrees away from the perpendicular—the poles will grow colder.

In the third cycle, lasting about 93,000 years, the earth's orbit becomes more elliptical, less nearly circular. This cycle influences the severity of an ice age resulting from seasonal differences in distance between sun and earth. And with the orbit now about halfway through the cycle and moving toward the more elliptical, it can be expected to intensify the effect of the 23,000-year cycle.

To test the Milankovitch theory, Hays and his co-workers studied sediment cores taken from sites midway between Africa, Australia and Antarctica where the sediment record could be expected to be sensitive to changes in climate.

Several analyses of the cores were made. By studying the remains of plankton—microscopic animals—that live in the ocean, the team could establish a record of water temperatures. Another set of measurements from the cores helped to establish a record of the amount of water stored in polar ice caps at various times. And in a third measurement, the abundance of a single Radiolaria species of plankton was used as an indicator of the presence of a specific set of water properties.

Two of the three analyses—the volume of the ice caps and water temperatures—consistently showed cycles correlating with the three orbital periods, while changes in the indicator of water properties provided a less reliable record of the two shorter orbital cycles.

Additional evidence in favor of the theory has since come from Ross Heath, Niklas Pisias and Ted Moore, all of the University of Rhode Island and CLIMAP project members. They studied sediment cores from three widely separated sites in the Pacific and found that changes in different measures of climate had periods similar to those predicted by the Milankovitch theory.

But is the theory fully explanatory? Hays, Imbrie and Shackleton had suggested that about 80 percent of the total variability observed in their sediment record was associated with the three cycles and "that changes in the earth's orbital geometry are the fundamental causes of the succession" of recent ice ages.

However, this conclusion has been questioned by several researchers, some of whom believe that the amount of variability explained by the three cycles is much less than 80 percent.

Climate cycles of even shorter periods than those predicted by Milankovitch are being identified, some much too short to be related to orbital variations. Some researchers suggest that long-term climate variations may be subject to control by a combination of Milankovitch cycles and other external controls and, as well, by regulating agents within the climate system.

And so, even though the Milankovitch theory and its validation thus far represent an important advance, here, as in so many other areas having to do with climate, much more research is needed.

And more research is clearly needed when it comes to human contributions to the complexities and problems of climate.

Some of those contributions, as we've noted in an earlier chapter, are clearly evident around cities: the increased temperatures because buildings and pavement store more heat than vegetation; the increased rainfall, thunderstorms and hailstorms (up by 430 percent near Houston).

But there are more human contributions.

Man's activities have been leading to a definite increase of carbon dioxide in the atmosphere—and carbon dioxide in high concentrations could change world climate.

An invisible and odorless gas, carbon dioxide (CO_2) consists of just one atom of carbon and two of oxygen and makes up a very tiny fraction, about 0.03 percent, of the atmosphere as contrasted with 20 percent oxygen and 78 percent nitrogen.

It's an important gas. For one thing, it is an important ingredient in the photosynthesis process by which green plants grow and produce oxygen. For another, carbon dioxide in the atmosphere allows through rays of sunlight which heat the earth's surface. When the heat is radiated back from the ground in the form of longer infrared rays, carbon dioxide tends to screen and absorb it, raising the temperature both of the gas and of the ground.

As the concentration of carbon dioxide is augmented, there is an increasing greenhouse effect that leads to greater warmth of the earth. And the higher the concentration of the gas, the greater the greenhouse effect and the warmer the earth may become.

There is no mystery about where the carbon dioxide is coming from. All fires—it makes no difference whether fueled by wood, coal, oil or gas—produce carbon dioxide, and millions of tons of it pour into the atmosphere every day. About half of it remains there. The rest may be soaked up by the oceans and by forests and other vegetation which use it for photosynthesis.

Since the Industrial Revolution began, huge quantities of carbon dioxide have poured into the atmosphere but only relatively recently, in the last quarter century or so, has the increase become of major concern.

In the last twenty years, according to measurements made by the Scripps Institution of Oceanography atop Mauna Loa volcano on the island of Hawaii—and confirmed by measurements at locations as varied as the South Pole, Alaska and Samoa—there has been as much of an increase in carbon-dioxide concentration as in the entire century before.

In two decades the concentration at the South Pole has increased from 314 to 331 parts per million. And it is estimated that since the beginning of the Industrial Revolution the rise has been about 13 percent.

While increased burning of fossil fuels is the major factor, many scientists worry about what they consider to be the perilous rate at which forests are being slashed and burned to extend agriculture and, especially in many developing countries, stripped to provide wood for fuel.

What are the likely future trends? For the remainder of this century, as *Science*, the Journal of the American Association for the Advancement of Science, sees it, the clearing of land will continue and the use of fossil fuels will increase. As a result, by the year 2000, the carbon-dioxide concentration will exceed preindustrial levels by about 25 percent. Ultimately, other forms of energy such as solar may play a more substantial role. But humanity's appetite for energy seems insatiable.

What will be the climatic consequences of increased carbon dioxide? Some investigators have calculated that a doubling of the gas could lead to an average global increase in temperature of 2.5 degrees Celsius.

While it should be noted that some scientists can be found who privately suggest that because of complex feedback phenomena, the net effect of increased carbon dioxide might be global cooling, the most likely trend, in the view of most scientists, appears to be warming. If so, the warming trend might counteract the cooling trend induced by the nonhuman factors.

But there is some concern that it might go beyond merely counteracting. If coal becomes the great energy hope of the country, and its use increases along the lines of some projections, then, according to a National Academy of Sciences study panel, atmospheric concentration of carbon dioxide might, by the year 2150, reach levels four to eight times those of the preindustrial era, producing an increase in the global mean air temperature of more than 6 degrees Celsius, or 11 degrees Fahrenheit, leading to climatic conditions not known since perhaps the age of the dinosaurs more than 70 million years ago.

Even a far smaller rise in temperature—of perhaps a degree or two—might have serious consequences. Enough ice might conceivably melt at the poles to raise sea levels by as much as 16 feet, inundating low-lying coastal areas around the world, including parts of our own Atlantic seaboard. There might well be heavier rainfall—enough possibly to restore fertility to the Sahara and the Arabian desert. But the wheat and corn belt of the central United States might suffer from excessive dryness.

No scientist is absolutely certain of any scenario for future climate. There are too many factors to be considered about which too little is known.

One factor is atmospheric dust which can act in either or both of two ways, cooling the earth by screening out sunlight but also keeping in heat by acting as an atmospheric lid. Dust impedes radiation in both directions, and top scientists are among the first to admit not knowing if the net effect is heating or cooling.

And ozone adds to the complexity. Ozone, (O_3) is one of the rare but vital gases in the atmosphere. There is only about one ozone molecule for every 2 million air molecules. The bluish gas consists of three atoms of oxygen and it is formed when sunlight strikes oxygen molecules.

Despite the small quantity of it, ozone is an efficient absorber of shortwave ultraviolet radiation and therefore acts as a filter. A reduction in atmospheric ozone would lead to an increase in solar radiation detrimental to humans, possibly with higher incidences of skin cancer as a result. And there could be other consequences.

A major part of the ozone is concentrated in a layer between 15 and 35 kilometers up in the stratosphere. The amount varies with time of day, season and latitude. There is more ozone during the day, for example, than at night; more over northern latitudes, less over southern (which has been correlated with the larger number of skin cancer cases in southern latitudes).

While ozone is produced by the action of sunlight on oxygen, the production is largely balanced by the reaction of ozone with nitrogen oxides and to a lesser extent by reactions with free chlorine atoms, which catalyze ozone destruction.

Increases in the amounts of nitrogen oxides and free chlorine would reduce the stratospheric ozone level. And for this reason, aircraft flying in the stratosphere (the SST problem) and use of fluorocarbons—the inert propellants in aerosol cans and re-frigerants—constitute environmental hazards that have caused public as well as scientific concern.

When, some years ago, it was announced that large fleets of supersonic transport planes were being planned, Dr. Harold Johnston of the University of California at Berkeley theorized that nitrogen oxide exhausts from the jets might lead to increased ozone destruction and cause a deficit.

Several years of study have confirmed the Johnston hypothesis, and by 1975 both the Department of Transportation and the National Academy of Sciences were warning that supersonic transport jet engines would have to be redesigned or flight of the planes would have to be limited.

By 1974, there was another worry when F. Sherwood Rowland and Mario Molina, chemists at the University of California at Irvine, pondered the environmental effects of fluorocarbons.

Earlier, some direct samplings by atmospheric chemists had shown increasing levels of fluorocarbons. Because the compounds are so resistant to change by other chemical entities in the lower atmosphere, Rowland and Molina determined that they must be floating up into the stratosphere unchanged. There, exposed to

164

strong sunlight, they would be dissociated, releasing chlorine atoms (each molecule in one of the most prevalent fluorocarbons has two chlorine atoms and there are three such atoms in the second most prevalent one). Chlorine would attack ozone. And, to make matters worse, the effects of fluorocarbon buildup could be expected to be delayed and long lasting because of the compounds' inertness and relatively slow ascent. The delay might be as long as ten years and the duration of effects as long as a century.

By 1976, there had been a sharp decline in the production of fluorocarbon propellants because of public awareness in the United States. But more than half the worldwide consumption is outside this country. And a new study released in 1979 concluded that the ozone in the atmosphere was being depleted twice as fast as had been estimated in a 1976 survey.

If the rest of the world continues to use fluorocarbons at the present rate, the ozone layer will be reduced by 16.5 percent. About half of this reduction of our protective ozone layer will take place in the next thirty years. The earlier report had estimated an ozone reduction of 7.5 percent. The use of improved computer programs and better data was the reason given for the latest findings.

A TWENTY-FIVE-YEAR FORECAST

Few climatologists agree on the weather picture even for the next half year let alone for any more extended period. But a recent study participated in by 24 climatologists in seven countries does provide a group prediction. Called "Climate Change to the Year 2000," the study was made by the National Defense University in Washington, D.C.

Although the study showed no consensus among the climatologists on any issue, it did indicate that the warming effect of carbon dioxide seems to be well established, and that the most probable outlook to the year 2000 is for a climate much like that of the last thirty years as a result of the balance of carbon dioxide warming and the cooling of the natural climate cycle.

From the responses of the experts, it appears that there is only one chance in ten of a global warming of more than 0.6 degrees centigrade above the temperatures of the early 1970s or of a cooling of more than 0.3 degrees centigrade.

Regardless of the temperature change they believe most likely, most of the climatologists predicted that any global temperature change will be magnified in the higher latitudes. But they believe that the "thermal inertia" of the Southern Hemisphere's ocans will reduce the magnification in that hemisphere. There also seemed to be a tendency to expect a twenty- to twenty-two-year drought cycle in the western United States.

In a nutshell, the outlook for the next twenty-five years is for the warming effects of atmospheric carbon dioxide to more or less balance the cooling cycle; for temperatures to increase uniformly in both hemispheres, with a slight warming trend in the north; for no significant change in the length or variability of the growing season.

Meanwhile, with more research clearly needed, we are about to get it. On the same day in 1978 when President Jimmy Carter achieved a plan for peace in the Middle East, he signed a document, perhaps humbler, but certainly more scientifically significant.

That document established the National Climate Program, which will be concerned with "anticipating the effects of climate fluctuations and changes in the United States and the rest of the world."

The program will coordinate United States climate research and funding which up to now has been spread through many agencies (for example, atmospheric research is carried on by at least ten agencies with a wide range of missions), and will stress climate monitoring, dissemination of data, assessment of human effects on climate, and impacts of climate on agriculture and the management of resources.

The program is expected to "fine tune" climate research, filling in gaps as well as coordinating and enhancing existing programs and starting new ones.

12

How Weather Affects You

THERE FREQUENTLY come into popular usage readings that are aimed at describing the weather's influence on our comfort. Many of the terms used to express these readings come and go— discomforture, sunburn index, frost factor, comfort index, effective comfort, etc.—but three seem to have captured public interest and remain in use in weather reports: temperature-humidity index (THI) in the summer and windchill factor and degree days in the winter.

First let's consider the THI, or temperature-humidity index. Various combinations of heat and moisture cause different sensations, and the National Weather Service adopted the term "temperature-humidity" to express this combined heat and moisture effect.

The formulae by which THI is determined are difficult. The equations are:

THI equals 0.4 (td + TW) + 15;
or THI equals .55td + .2tdp + 17.5;
or THI equals td − (0.55 − 0.55RH) × (td − 58);

where td is the temperature, TW is the wet-bulb temperature, tdp is the dew-point temperature, RH is the relative humidity.

According to the National Weather Service, which issues the reading, relatively few people will be made uncomfortable by summer heat and humidity when the THI is 70 or below. About half the people become uncomfortable when the THI reaches 75 and almost everyone is uncomfortable when the THI climbs to 79. Over 80 the THI would indicate acute discomfort. One must also remember that air movement plays an important role in comfort, and the THI does not take this element into consideration.

167

It is easier to refer to a table to get the THI than to resort to the formulae.

Temperature-Humidity Index Table
(derived from *Sums* of Simultaneous
Dry- and Wet-Bulb Temperatures)

Sum	THI	Sum	THI	Sum	THI	Sum	THI	Sum	THI
100	55	120	63	140	71	160	79	180	87
101	55	121	63	141	71	161	79	181	87
102	56	122	64	142	72	162	80	182	88
103	56	123	64	143	72	163	80	183	88
104	57	124	65	144	73	164	81	184	89
105	57	125	65	145	73	165	81	185	89
106	57	126	65	146	73	166	81	186	89
107	58	127	66	147	74	167	82	187	90
108	58	128	66	148	74	168	82	188	90
109	59	129	67	149	75	169	83	189	91
110	59	130	67	150	75	170	83	190	91
111	59	131	67	151	75	171	83	191	91
112	60	132	68	152	76	172	84	192	92
113	60	133	68	153	76	173	84	193	92
114	61	134	69	154	77	174	85	194	93
115	61	135	69	155	77	175	85	195	93
116	61	136	69	156	77	176	85	196	93
117	62	137	70	157	78	177	86	197	94
118	62	138	70	158	78	178	86	198	94
119	63	139	71	159	79	179	87	199	95

The second comfort index, frequently included in weather reports during the cold months, is the windchill factor. A reading developed by the Army Quartermaster Corps about forty years ago, it is a determination of the effect of the combination of wind and temperature upon exposed skin surfaces. The formula for this is even more complicated than those for determining the THI, but

WINDCHILL TABLE

°F. DRY-BULB TEMPERATURE

WINDCHILL INDEX

(EQUIVALENT TEMPERATURE—Equivalent in cooling power on exposed flesh under calm conditions)

MPH	35	30	25	20	15	10	5	0	-5	-10	-15	-20	-25	-30	-35	-40	-45
calm	35	30	25	20	15	10	5	0	-5	-10	-15	-20	-25	-30	-35	-40	-45
5	33	27	21	16	12	7	1	-6	-11	-15	-20	-26	-31	-35	-41	-47	-54
10	21	16	9	2	-2	-9	-15	-22	-27	-31	-38	-45	-52	-58	-64	-70	-77
15	16	11	1	-6	-11	-18	-25	-33	-40	-45	-51	-60	-65	-70	-78	-85	-90
20	12	3	-4	-9	-17	-24	-32	-40	-46	-52	-60	-68	-76	-81	-88	-96	-103
25	7	0	-7	-15	-22	-29	-37	-45	-52	-58	-67	-75	-83	-89	-96	-104	-112
30	5	-2	-11	-18	-26	-33	-41	-49	-56	-63	-70	-78	-87	-94	-101	-109	-117
35	3	-4	-13	-20	-27	-35	-43	-52	-60	-67	-72	-83	-90	-98	-105	-113	-123
40	1	-4	-15	-22	-29	-36	-45	-54	-62	-69	-76	-87	-94	-101	-107	-116	-128
45	1	-6	-17	-24	-31	-38	-46	-54	-63	-70	-78	-87	-94	-101	-108	-118	-128
50	0	-7	-17	-24	-31	-38	-47	-56	-63	-70	-79	-88	-96	-103	-110	-120	-128

VERY COLD

BITTERLY COLD

EXTREME COLD

Wind speeds greater than 40 mph have little additional chilling effect.

you can get the windchill factor, or index, by finding the temperature on the horizontal line at the top of the chart on page 169 and then the average wind speed in the leftmost vertical column, and where these parameters intersect, you can read the windchill.

For example a temperature of 10 degrees Fahrenheit and a wind of 20 mph is equivalent to a windchill of −24 degrees Fahrenheit.

The third reading often issued by the National Weather Service is called the "degree day." It has caused untold confusion among the lay audience because the degree day concept was originally devised by and for heating engineers. The primary users of heating degree days are fuel oil and coal distributors, public utilities and users of large quantities of heating fuels. Now the individual home owner is making use of the concept. Let me explain how it works.

Heating engineers have determined that to maintain a comfortable indoor temperature of 70 degrees Fahrenheit, it is usually necessary to use the indoor heating plant whenever the outdoor temperature drops below 65 degrees. So 65 is used as a base number. When the average temperature during the day (the high plus the low divided by 2) drops below 65, that average temperature is subtracted from the base figure of 65. For example: when the high is 60 degrees and the low is 50, then the average reading for the day is divided by 2, or 55 degrees. Subtracting 55 from the base 65 gives 10 degree days.

National Weather Service offices throughout the country maintain daily records of the degree-day figures. They are also kept on a cumulative basis so that one can compare the number of degree days during one season with another and in this way determine which season was colder.

More than 2,000 years ago, in the fifth century B.C., Hippocrates, the father of medicine, wrote a treatise, *Airs, Waters and Places,* in which he emphasized the effects of weather on disease.

Solemnly, he urged his young physician disciples not to ignore meteorology, a science which he held to be closely allied to medicine. The first thing a young physician setting up his practice should do, Hippocrates emphasized, should be to study the local climate, the seasons, the weather conditions, "for with the seasons, men's diseases, like their digestive organs, suffer change."

Among other things, Hippocrates considered dry years to be healthier than wet; he asserted that rainy seasons promoted protracted fever, epilepsy, apoplexy and quinsy. On the other hand, he did allow that in dry weather, arthritic ailments, consumption, dysentery, and eye inflammations are prevalent. North winds were connected with coughs, throat ailments, constipation and pains; south winds with dull hearing, dim vision, weakness and inactivity, and also with loosening of the bowels, headaches and vertigo. But, he noted, it was true that some people, because of differing temperament and body build, have just the opposite reactions to weather changes as do others.

Up until the nineteenth century, most efforts to explain diseases revolved about weather and environment. At that point came the discovery of germs, and medicine concentrated on bacteriology and related sciences.

After a dormant period, however, the study of the effects of weather and climate on human health, physical and emotional, has flowered again.

As a science, biometeorology, as it is called, is new, very young, quite tentative. But in the last several decades, it has begun to come in for increasing attention from physicians, physiologists and meteorologists.

Notably in France, there have been efforts to make an exact science, called *meteorolopathologie,* out of studies of adverse human reactions to weather. And some years ago, in West Germany, a service was set up for physicians who could phone a private number to receive a special weather forecast intended to help them in their treatment of patients. Typically, a physician-caller might be told: "Tomorrow, we look for penetration of humid warm air masses from the Mediterranean, and it is advisable to pay special attention to patients with circulatory ailments." Reportedly, German surgeons in particular have been concerned

171

with weather conditions on days when they operate, especially when operations are delicate and could be postponed if conditions are unfavorable—when, for example, hot and humid weather might be conducive to hemorrhage.

Some weather effects are fairly obvious, among them, for example, the increase in the common cold with the changeable weather of early spring. But there have been probings in recent years of many other less apparent correlations between aspects of weather and human activities.

Among those who looked at the kinds of bodily changes weather can produce was Dr. Clarence A. Mills, who was director of the University of Cincinnati's Laboratory of Experimental Medicine. And one of the most dramatic changes, he reported, is the swelling and unswelling of tissues with shifts of barometric pressure.

What happens if you hold a sponge tightly squeezed in your hand in a dish of water and then relax the pressure? The sponge of course absorbs water. Increase the pressure, and the water is released.

In much the same way, the studies of Mills suggested, body tissues sop up more water from the intestinal tract when atmospheric pressure falls—enough, for example, to add an inch to leg girth over a twenty-four-hour period.

Mills also noted that tissue swelling increases pressure within the brain. And, because the brain is tightly encased in the rigid skull, the swelling may squeeze blood vessels, diminishing blood flow. This, Mills held, could lead to despondency, irritability and loss of mental acuity. Let the barometer rise again and the extra fluid is squeezed out of the tissues, and life becomes rosier and mental acuity increases.

Temperature, too, can have significant effects. The body accommodates to hot and cold weather by increasing or decreasing the activity of sweat glands and by some shifting of blood into or away from the skin. But when heat is prolonged, further adjustments are needed. The thyroid, adrenals and other glands become less active to bring about a reduction of metabolism, or internal combustion, so less body heat is produced. But this can mean less energy for thinking and acting.

Most people manage to get through such weather-induced distress. But, it has been held, for people whose physical condition is under par to begin with, the effects of bad weather can conceivably turn marginal health into illness.

The late Dr. Ellsworth Huntington of Yale, another prominent investigator, observed that Denver schoolchildren had to be disciplined five times more often in humid than in dry weather.

Huntington also noted that of 148 religious riots occurring between 1919 and 1941, more than one-third took place in just two months, April and August, the most uncomfortable there.

In New York City, another investigator, O. E. Dexter, studied the relationship between weather and assault-and-battery arrests. Looking at 40,000 such cases, Dexter found that the arrest rate increased as temperature increased. From a low in January, it rose to a peak in July, falling off, however, during the heat and humidity of August.

Dexter's conclusion was that "Temperature, more than any other condition, affects the emotional states which are conducive to fighting."

On a much more plebeian level, spring fever is common around mid-April, especially when a sudden warm spell follows a long cold period. There is a feeling of lassitude that appears to be related to the body's efforts to accommodate to the change in weather. In the body's efforts to get rid of heat, more blood is carried to vessels near the skin. And it has been held that in the early process of the change in blood circulation, blood plasma, the watery substance of the blood, may increase in amount, accounting for that old belief that the blood "thins" in spring. The feeling of lassitude may be explained by the large amount of work involved in the body's shifting of blood circulation. After a few days of acclimatization, spring fever disappears.

Steady, monotonous winds that blow for long periods have long been believed to have profound influence on people. Many North Africans are convinced that the dusty, hot sirocco wind blowing off the Sahara can depress many people even to the point of suicide. In Tangiers, the levanter east wind from the Mediterranean is blamed for headaches and oppressive feelings. The warm, moist *vent du midi* wind in southern France is often connected popularly with asthma attacks, headaches, rheumatic pains and epileptic seizures.

Cyclonic storms also have been held to be capable of producing significant mental and physical changes. Dr. Mills has described the frequent cyclonic storms in the midwest as "leaving behind them a trail of human wreckage—cases of acute appendicitis, respiratory attacks of all kinds, and suicides."

According to Dr. Mills, even healthy people may be affected by such storms. As the barometer falls and humidity rises with the approach of the storm front, people are often bothered by "a feeling of futility, an inability to reach the usual mental efficiency or to accomplish difficult tasks." Children become irritable, adults quarrelsome. "Such weather," Mills has said, "provides the most perfect background for marital outbursts."

When the storm front passes and the barometer rises and humidity falls, people's spirits do an about face and become buoyant.

Some diseases have long been credited with acting like weather vanes. Dogs have been said to be able to smell a storm coming. Some people seem to be able to do the same.

Among them are older people, the allergic and the chronically ill who will tell you that they feel coming storms in their bones, joints, muscles, sinuses and heart rhythms. Both gout and sciatica are often more painful before a storm. And there have been some studies that suggest that changes in pulse rates, breathing rates, blood pressure and other physical processes may be related to the passing of low- and high-pressure air masses that precede and follow storms.

Although long regarded as a kind of old wives' tale, the weather sensitivity of arthritics—their ability to predict changes because of increased pain at such times—has received scientific support.

174

At the University of Pennsylvania Hospital in Philadelphia, arthritic patients volunteered to live for periods of two to four weeks in a climate chamber where, without their knowledge, changes could be made in temperature, pressure, humidity and air movement.

When only a single factor was changed, none of the thirty patients experienced any change in symptoms. But when investigators tried duplicating approaching storm conditions—gradually lowering barometric pressures from 31.5 to 28.5 inches and increasing humidity from 25 to 80 percent—most of the patients complained, within just a few minutes, of increased pain and their joints became stiff and swollen.

Does weather ever actually cause disease? While there is no evidence that it does so directly, many investigators believe that it may weaken body resistance to disease by modifying mental outlook and physical processes.

Chilling, for example, may weaken body resistance to many diseases. Even in the tropics, there are indications that outbreaks of sickness often follow sudden decreases in air temperature. Very low humidity, especially when coupled with high winds, may fill the air with dust which can irritate breathing passages and may increase susceptibility to infection.

In a series of studies that I undertook at the Albert Einstein College of Medicine, it was found that there were sometimes marked increases in emergency-room visits for asthma attacks at several major city hospitals. The increased numbers of such visits were significantly correlated with the first major cold spells in the early fall. Such statistical studies can pinpoint the relationship but not the cause. The study, which was published in the *Archives of Environmental Health*, concluded:

> The most pronounced and statistically significant increase in asthma visits occurred with the onset of the first or second cold periods, that is, periods requiring heating. These significant increases in asthma visits do not

175

appear to be related to pollens, molds, or the levels of sulfur dioxide, smoke shade, or carbon monoxide in the atmosphere.

The relationship between increased asthma visits and the onset of cold weather may be a geographically widespread phenomenon.

Few statistical studies which deal with the incidence of asthma and the possible causes for sudden asthma outbreaks are available. Epidemiological investigations should be encouraged to provide this much needed information. Emergency clinic visits are an excellent source of data for such studies.

According to some studies, migraine, colic, stroke and epileptic seizures seem to occur more often in cool, damp weather, and lung clots, vein clots and hemorrhage occur predominantly in warm, damp weather.

Upper respiratory infections, including the common cold, are of course more common in winter than summer, four-and-a-half times more frequent in January, one of the peak months, than in July. Actually, the organisms that produce the infections are prevalent year-round. But recent studies indicate that viruses can grow and multiply only within a restricted temperature range. For many that infect humans, the optimum growing temperature appears to fall about a degree or two below normal human body temperature. Cold weather might lower body temperature just enough to permit multiplication of some previously inactive viruses. On the other hand, a rise of body temperature may help check virus growth, which could be the reason why a fever may play some part in combating infection.

There have been investigations indicating that asthma patients sometimes experience increased symptoms during sudden temperature drops. Some studies have pointed to an increase in mental hospital admissions when there is excessive electrical activity in the atmosphere associated with lightning and sunspots.

The relationship between weather and the heart, blood pressure and blood has also come in for investigation. Some years ago, physicians in Holland in the course of making routine blood pressure checks of normal, healthy Red Cross blood donors were puzzled to find that the same donors sometimes had pressures 20 percent above their usual pressures. Then a pattern began to emerge. It was during January and February that 77 percent of the donors showed elevated pressures. Later, additional studies of healthy people indicated that blood pressure tends to be higher in cold than in warm weather.

Changes in blood have also been noted. In winter, hemoglobin—the pigment that gives red blood cells their color and also serves to transport oxygen—tends to be less. And there is also less gamma globulin, the blood fraction containing infection-fighting antibodies. There have also been reports from German and Japanese scientists of lower blood values for calcium, magnesium and phosphates in winter, and from Swiss workers, of increased resistance in the capillaries, the very tiny blood vessels, during cold weather. It is possible that such findings are related to health problems.

The heart can be affected by extremes of weather. In very cold weather, it must work harder to maintain normal body temperature, and even healthy people may feel some strain and a tendency to tire more easily. Very hot weather increases blood flow in the body and puts more strain on the heart.

Dr. George E. Burch, a distinguished cardiologist at Tulane University School of Medicine, has studied climate effects on the heart for more than twenty years and has found extreme heat or cold to be especially critical for heart patients, with pronounced strain occurring during the first days of either extreme as the heart and body adjust to new temperature conditions.

In summer months in subtropical United States cities heart attack deaths increase. In those northern cities where summers are relatively mild but cold weather is severe, studies by Burch and his colleagues indicate that the death rate is highest in winter.

But, as the Tulane investigators point out, three-fourths of the United States has hot and humid summer weather and deaths from heart conditions increase significantly in these regions, especially during heat waves.

Burch and others have found that almost two-thirds of patients who were being treated successfully for congestive heart failure quickly developed failure again in hot, humid surroundings—and in every case the failure responded promptly to treatment again when the patients were placed in a comfortable environment.

Out of such studies have come useful suggestions for people with heart problems. They should realize the hazards of high heat and humidity and as much as possible avoid outdoor activity or limit it sharply during hottest times of day. For the severely disabled, air conditioning may be as important as good diet and medical care. When air conditioning is not available, it can be helpful to wear loose, light clothing, and to sponge the skin frequently with tepid water to help the body lose heat.

Still other effects of weather on our physical well-being are continually undergoing examination.

Studies by Dr. E. W. Hartman of the Margaret Sanger Research Bureau indicate that heat tends to inhibit sperm production, and there is a seasonal rhythm of male fertility, with a significant drop in the conception rate in the South during the summer—by as much as 30 percent in Florida, for example.

There have been studies, too, suggesting that hot summer weather may affect the intelligence of children born the following winter. One theory holds that, as important fetal brain areas are formed during the third month of pregnancy, any damage to the fetus at that time may affect IQ, and such damage may occur if the mother significantly reduces her food intake, especially intake of protein, which she may be more likely to do in hot weather than at other times. One study of admissions of the mentally impaired to an Ohio institution over a period of many years found that a high percentage had been born in the first three months of the year so that for them the third month of fetal life occurred during summer.

According to one British study, whether or not rheumatoid arthritis develops in later life may be determined at least in part, somehow, by season of birth. The incidence of the disease was found to be higher among people born between September and February than among others born between March and August.

Surveys indicate that, considered in terms of rates of absence from work because of illness, June is the most healthful month of the year, February the least healthful. On an average February day, according to one study, 2.65 percent of all workers are at home sick.

Still other studies have indicated that eye and eyelid inflammations are highest in April and lowest in October; middle ear infections peak in February and are lowest in August; flu is highest in February, lowest in July; sore throat and tonsillitis are highest in January and lowest in July; and sinusitis as well as the common cold is highest in February and lowest in July.

There also have been efforts, of course, to try to determine how weather may affect the mind and its workings. According to investigations by Dr. Huntington, while physical performance is best at 64 degrees Fahrenheit, for mental function the best temperature is in the range of 38 to 40 degrees Fahrenheit. College students taking standard intelligence tests have generally done only about 60 percent as well in summer heat as in winter cold.

Animal studies also point to a strong influence of temperature on mental function. At 55 degrees Fahrenheit, it took rats twelve tries before they found the right path through a maze to a food dish. At 75 degrees, other comparable rats needed twenty-eight tries; and others, put to the test at 90 degrees, needed fifty. Moreover, a month later, tested again in the maze—this time for memory—the 55-degree rats knew the right path immediately; the 75-degree animals had to relearn half the steps; the 90-degree rats exhibited no memory of the maze at all.

Temperature also seems to have effects on hearing. According to University of North Carolina studies, hearing may be keenest at 50 degrees Fahrenheit, with acuity somewhat reduced as temperature goes above or below that point.

Even fog may have effects on mental functioning, with some

decrease in efficiency during foggy weather as contrasted to fair. At one time, the Bank of England had a rule requiring all important files to be locked up on excessively foggy days because at such times clerks keeping the records made a much higher percentage of errors.

An unusual—and unintentional—experiment took place at Massachusetts State College at Amherst in September 1938 at the time of the famed New England hurricane. The entering freshman class was taking three standard tests, one on the day before the hurricane, the second during passage of the hurricane and the third on the day afterward. In the first test, results were slightly above average; in the last test, a bit below. But the differences were not significant. However, the scores for the test taken during passage of the hurricane—when the sky was dark, the winds howled, trees were crashing to the ground, and the pressure first dropping rapidly and then rising abruptly—were 20 percent above those usual at the school. Apparently, the hurricane in some way had acted as a powerful stimulant.

The seeming ability of weather to affect mood is known to many, who attribute feelings of excitement, "blueness," laziness, ambition, nervousness or calm to what is going on with the weather.

Barometric changes in particular have come in for study. It appears that low barometric pressure may cause restlessness and interfere with concentration, make adults frustrated and quarrelsome, and children irritable. Police records in large cities indicate that more acts of violence, including suicides, tend to occur when barometric pressure goes below 30.00 inches. In one large industrial plant, records over a fifty-year period indicate that 74 percent of all lost-time accidents occurred when pressure was below 30.00.

Some years ago, writing in an American Medical Association publication, Dr. Noah Fabricant pointed out that Abraham Lincoln was subject to episodes of mental depression that disappeared when barometric pressure became stable. Wrote Fabricant: "He and his wife, Mary Todd Lincoln, might have been spared a number of crises in their personal lives had they fully understood the effect of the weather."

Dr. Clarence A. Mills, too, advised considering the barometer when moods change. "If," he wrote, "yesterday's bright idea seems pretty poor today, check the barometer. Knowledge that weather may be the basis of your blues as well as boosting you to your emotional peak often can be of great help in achieving a more tranquil existence. Blame your own and the other person's bad moods on the weather—and rest assured that a change is just around the corner."

Science is still far from knowing the answer to the question of how environmental factors exert their influences. One clue, some investigators believe, may lie in ions—minute, electrically charged particles found in the atmosphere.

There are both positively and negatively charged ions and they are usually found in the air in a ratio of five positive to four negative. For many years, ionization investigators have speculated that that ratio is important, and that if it is upset and the number of positive ions is increased, there might be undesirable effects on the body. Whereas negative ions are partly composed of oxygen, which is beneficial to the body, positive ions are partly composed of carbon dioxide which can be harmful.

In one study at the New York University College of Engineering, volunteers were exposed to negative ions and given tests. The results indicated that their visual responses had improved and they could work harder, as well, without obvious fatigue. At the University of California, in similar studies, Dr. Albert P. Krueger has found that an excess of negative ions leads to better functioning of the respiratory system.

What can produce an imbalance in favor of positive ions? Studies indicate that smog and some kinds of heating and air-conditioning equipment can do so. Some investigators have suggested that an imbalance may stem from atmospheric conditions—that, for example, there may be an excess of positive ions in the air just before a storm but when the storm breaks, it may introduce more negative ions.

As you might expect, since people vary in many other ways, they also vary in their responses to weather. Generally, the healthy are much less affected by weather changes than those in

poor health. It also appears that women generally—with exceptions of course—endure weather extremes better than men and
are particularly superior in adapting to cold. That may be because
women on the average are smaller in stature and have less surface
area and tend to have more fatty-tissue insulation, so their bodies
may conserve heat more efficiently.

Dr. Huntington's studies suggest that reactions to changes in
weather can be influenced by where one lives. Someone spending
most of his life in the relatively stable climate of San Francisco, for
example, Huntington found, reacts seventeen times more strongly
to weather changes than a person living in the stormy region
around Minneapolis. Apparently people used to abrupt and fierce
weather swings become adapted and are better able to take them
in stride.

Is there anything one can do about weather effects?

Some things, yes. If you're generally healthy, but the weather
gets you down on occasion, it could be a good idea to keep a record
for a time to find out exactly what kind of weather is responsible.
Even if you can't avoid oppressive weather, you can take it into
account. Aware that you're in for an off day or two—and why—you
can plan to do less- rather than more-important jobs, avoid major
decisions, avoid frustration, knowing you'll feel better the next day
or the day after.

You might want to check, too, on weather influences on your
spouse and children. Many of their dark moods, you may find to
your relief, are merely responses to weather and nothing more.

And you might want to take into account some of the findings of
Dr. Huntington: that generally people tend to be at their best in
terms of both health and energy in November and at their worst in
January, February and March, when they tend to be more tired
and susceptible to disease. It's during this period of tiredness that
a vacation may make the most sense—although it's at this time
that people, rather than vacationing, often make resolutions to
work even harder. According to Huntington, "To speed up at the
end of January is like taking a tired horse and expecting him to win
a race."

If you have a health problem that might be influenced by weather, one for which possibly another climate might be helpful, medical advice is in order.

Any move to a different climate for health reasons needs very careful consideration. Other factors can be very important. Some arthritic patients, for example, may find some measure of relief in warm, dry areas. But it is possible that no small part of that relief stems from the relaxation, physical and mental, that often goes along with a trip to such an area. A permanent move, however, may be another story. Uprooting one's self and family, moving away from long-time friends may have emotional repercussions that might counteract the climate and nullify its effects or even make the arthritic problem worse.

If you're thinking of making a move, go slowly. Usually, the health problems that may benefit from a climate change are chronic, long-term, slow-changing ones. There is no need for a hasty decision.

Review the situation with your physician and perhaps even with specialists he can suggest. For one thing, it is a fact that new treatments for many problems are constantly being developed. It's possible that a specialist can suggest a trial of a new treatment that may promise as much relief as a change of climate.

If your family physician and specialists agree that the change of climate is needed, you can act in gradual steps. Give yourself a trial period. Try the new area for a few months, if possible, in both winter and summer. If you feel fine all year, good. On the other hand, you may find that the beneficial effects of a dry climate in winter may be counterbalanced or even outweighed by the debilitating effects of blistering summer heat. Only actual experience can give you any assurance of what to really expect.

Tell Me, Please:
The Questions Viewers Often Ask

What accounts for the seasons?

In a nutshell, the earth's leaning, or tilt.

The earth's axis is actually tilted about 23½ degrees from upright position—that is, perpendicular to the plane of the earth's orbit. This leaning is constant throughout the daily rotation of the earth on its axis and also throughout the annual revolution around the sun.

Since the tilt remains the same, at times the North Pole will be leaning toward and at other times away from the sun. When the North Pole leans toward the sun, the sun's rays strike the Northern Hemisphere in a more direct and concentrated manner, bringing summer to the northern half of the earth. At the same time, the South Pole is angled away from the sun, with winter conditions then prevailing over the Southern Hemisphere.

When is the earth nearest to and farthest from the sun?

You may be a bit astonished by the answer to this.

While the mean distance of the earth from the sun is 93 million miles, the earth's orbit around the sun is an ellipse rather than a circle. At the beginning of January, the earth is actually about 1.6 million miles nearer the sun than the mean distance, and at the beginning of July, the same amount farther away.

Although you might think that in January, with winter in the Northern Hemisphere, the earth must be farthest from the sun, that isn't true. The cold of winter comes because the sun is low in the sky and solar energy spreads over a larger area. For example, when the sun is 45 degrees above the horizon, the area of spread is 22 percent larger than when it is 60 degrees above. The rays then have a longer path through the atmosphere and lose more energy on the way to the earth's surface.

What are the solstices and equinoxes?

Solstice refers to a position of the sun in the sky with relation to the equator of the earth. When, as the earth revolves around the sun, the sun appears farthest south of the equator, we have the winter solstice, or beginning of winter. When the sun reaches a point farthest north of the equator, we have the summer solstice. These are the solstices for the Northern Hemisphere. The Southern Hemisphere experiences exactly the opposite solstices.

The equinox is a point midway between two solstices. For the Northern Hemisphere, the vernal equinox, marking the beginning of spring, occurs at the moment when the sun moves across the equator on what appears to be its journey northward. The autumnal equinox occurs when the sun appears to cross the equator heading south.

Usually, the winter solstice occurs on December 21, the vernal equinox on March 21, the summer solstice on June 21, and the autumnal equinox on September 22. The dates may vary a day either way.

How much energy do we get from the sun?

The sun is not beaming its energy toward the earth alone but in all directions, and the earth intercepts only a small fraction of it. Although this is a vast amount—170×10^{12} kilowatts, or more than 500,000 times the capacity of all the electricity generating plants in the United States—the total energy emitted by the sun is more than 2 billion times that much.

How long before the sun burns itself out?

The sun is a huge nuclear reactor. Under the high-temperature and pressure conditions in the interior of the sun, a hydrogen-helium fusion process takes place, transforming mass into energy.

But the amount of mass used is so small, it will take about 4.6×10^{12} years for the mass to be exhausted. The estimated age of the solar system is 4.6×10^{9} years, so the sun can be expected to last more than 3,000 times as long as it has existed thus far.

Three scales are used for temperature in meteorology. How can one be converted to another?

Fahrenheit, of course, is the oldest, dating back to the beginning of the eighteenth century. It considers that at sea level water has a boiling point of 212 degrees and a freezing point of 32 degrees. These numbers are not very logical, nevertheless the Fahrenheit scale has been used long and widely.

Centigrade, also know as Celsius, was devised in the mid-eighteenth century by Anders Celsius, a Swedish astronomer. It considers that a mixture of ice and water has a temperature of 0 degrees and boiling water has a temperature of 100 degrees.

The third scale, the absolute or Kelvin, named after the English scientist, Lord Kelvin, who suggested it, takes zero to be absolute zero, the lowest temperature that, theoretically, can exist, a point at which molecules would stop vibrating. On the Celsius scale, absolute, or Kelvin, zero corresponds to -273.18 degrees; on the Fahrenheit, to -459.72 degrees. And, in fact, scientists have been able to achieve temperatures close to absolute zero.

Converting Celsius to Fahrenheit requires simply multiplying the Celsius temperature by 1.8 and adding 32 degrees.

For the reverse conversion, from Fahrenheit to Celsius, subtract 32 degrees and divide by 1.8.

Why do weather reports include more and more metric terms?

The National Weather Service is developing plans for converting to metric public weather forecasts as soon as the nation is

THINK METRIC
TEMPERATURE

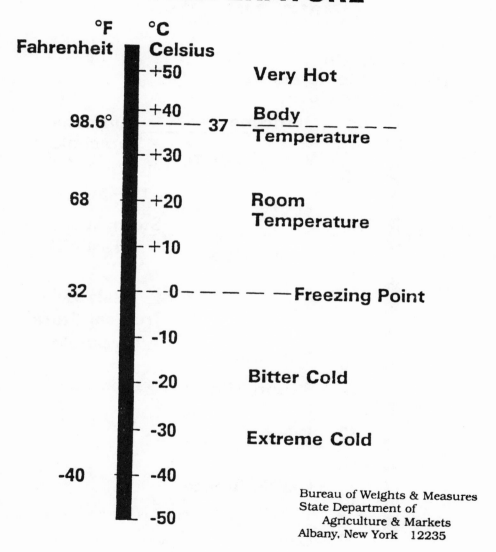

°F
Fahrenheit

°C
Celsius

+50 — Very Hot

98.6° — +40 — 37 — Body Temperature

+30

68 — +20 — Room Temperature

+10

32 — -0 — Freezing Point

-10

-20 — Bitter Cold

-30 — Extreme Cold

-40 — -40

-50

Bureau of Weights & Measures
State Department of
Agriculture & Markets
Albany, New York   12235

THINK METRIC
WIND SPEEDS

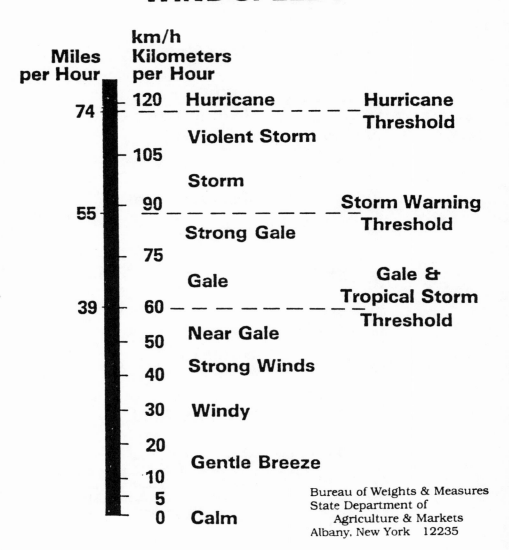

Miles per Hour

km/h Kilometers per Hour

Miles per Hour	km/h		
74	120	Hurricane	Hurricane Threshold
		Violent Storm	
	105		
		Storm	
55	90		Storm Warning Threshold
		Strong Gale	
	75		
		Gale	Gale & Tropical Storm Threshold
39	60		
	50	Near Gale	
	40	Strong Winds	
	30	Windy	
	20		
		Gentle Breeze	
	10		
	5		
	0	Calm	

Bureau of Weights & Measures
State Department of
Agriculture & Markets
Albany, New York 12235

THINK METRIC
PRESSURE

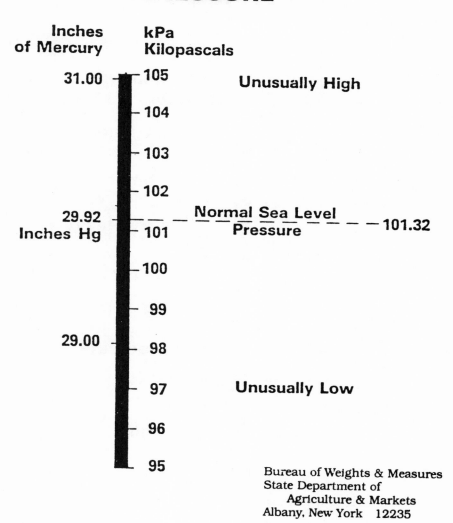

Inches
of Mercury

kPa
Kilopascals

31.00 — 105   Unusually High

— 104

— 103

— 102

29.92 — — — Normal Sea Level — — — 101.32
Inches Hg   101   Pressure

— 100

— 99

29.00 — 98

— 97   Unusually Low

— 96

95   Bureau of Weights & Measures
State Department of
    Agriculture & Markets
Albany, New York   12235

THINK METRIC
RAINFALL RATE

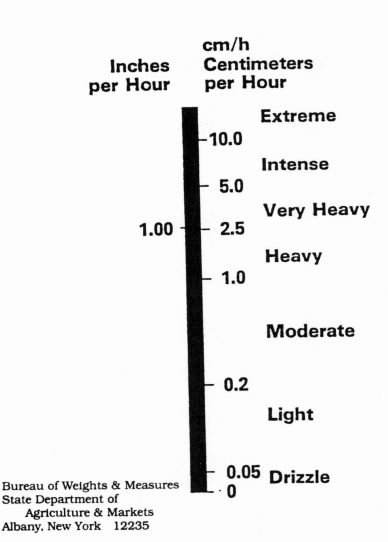

Inches per Hour

cm/h
Centimeters per Hour

Extreme

10.0

Intense

5.0

Very Heavy

1.00 — 2.5

Heavy

1.0

Moderate

0.2

Light

0.05 Drizzle
0

Bureau of Weights & Measures
State Department of
    Agriculture & Markets
Albany, New York   12235

THINK METRIC
SNOWFALL

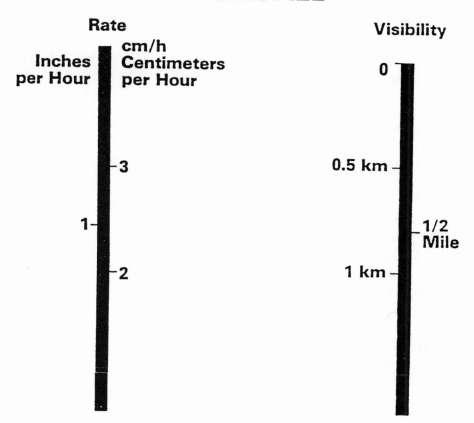

Rate

Inches per Hour

cm/h
Centimeters per Hour

3

1

2

Visibility

0

0.5 km

1/2 Mile

1 km

Heavy Snow Warnings Are Issued When:
- **10 cm or more in 12 hours is expected.**
- **15 cm or more in 24 hours is expected.**

Bureau of Weights & Measures
State Department of
    Agriculture & Markets
Albany, New York   12235

ready. Most countries, including Mexico and Canada, are on the metric system already. In addition to the degree Celsius, metric units used in meteorology are the kilopascal (kPa) for air pressure, kilometers per hour (km/h) for wind speed, millimeters (mm) for rainfall and centimeters (cm) for snow depth.

How fast can temperatures drop in cold waves?

One of the fastest falls recorded occurred in Rapid City, South Dakota, on January 12, 1911, when the temperature plummeted from 49 degrees Fahrenheit at 6:00 A.M. to −13 degrees at 8:00 A.M., a 62-degree fall in two hours.

Later that same year—in November—Kansas City, Missouri had a drop from 76 degrees Fahrenheit at 10:30 A.M. to 10 degrees at midnight and to 7 degrees Fahrenheit at 7:00 A.M. the next morning.

What's the worldwide temperature range?

Temperatures measured at a level of 5 feet in the air range from 125 degrees Fahrenheit below zero in Antarctica to more than 135 degrees Fahrenheit above on the Libyan desert—a spread of 260 degrees.

Where else in the world have record high and low temperatures been recorded?

HIGHEST (in degrees Celsius)

Africa	57.8	Azizia, Libya	9/12/22
N. America	56.7	Death Valley, CA	7/10/13
Asia	53.9	Tirat Tsvi, Israel	6/21/42
Australia	53.3	Cloncurry, Queensland	1/16/1889
Europe	50.0	Seville, Spain	8/4/1881
S. America	48.9	Rivadavia, Argentina	12/11/05
Oceania	42.4	Tuguegarao, Philippines	4/29/12
Antarctica	14.4	Esperanza, Palmer Peninsula	10/20/56

Antarctica	-88.3	Vostok	8/24/60
Asia	-67.8	Oimyakon, USSR	2/6/33
Greenland	-66.1	Northice	1/9/54
N. America	-62.8	Snag, Yukon	2/3/47
Europe	-55.0	Ust'Shchugor, USSR	exact date unknown
S. America	-32.8	Sarmiento, Argentina	1/1/07
Africa	-23.9	Ifrane, Morocco	2/11/35
Australia	-22.2	Charlotte Pass NSW	1/22/47
Oceania	-10.0	Haleakala Summit, Maui	1/2/61

What are the extremes of hot and cold for the fifty states?

	COLD (in degrees Fahrenheit)	HOT (in degrees Fahrenheit)
Alabama	-18	112
Alaska	-76	100
Arizona	-37	127
Arkansas	-29	120
California	-45	134
Colorado	-60	118
Connecticut	-32	105
Delaware	-17	110
Florida	- 2	109
Georgia	-17	112
Hawaii	14	100
Idaho	-60	118
Illinois	-35	117
Indiana	-35	116
Iowa	-47	118
Kansas	-40	121
Kentucky	-34	114
Louisiana	-16	114
Maine	-48	105
Maryland	-17	109

Massachusetts	-32	102
Michigan	-51	112
Minnesota	-59	114
Mississippi	-16	115
Missouri	-40	118
Montana	-70	117
Nebraska	-47	118
Nevada	-50	122
New Hampshire	-46	106
New Jersey	-34	110
New Mexico	-50	116
New York	-52	108
North Carolina	-23	109
North Dakota	-60	121
Ohio	-39	113
Oklahoma	-27	120
Oregon	-54	119
Pennsylvania	-42	111
Rhode Island	-23	105
South Carolina	-13	111
South Dakota	-58	120
Tennessee	-32	114
Texas	-23	120
Utah	-50	118
Vermont	-50	105
Virginia	-29	110
Washington	-42	118
West Virginia	-37	112
Wisconsin	-54	114
Wyoming	-66	114

What are the differences between conduction, convection and radiation?

Conduction is the transfer of heat by direct contact—as when your hand picks up heat from a hot surface. Because air is a poor conductor, conduction insofar as weather is concerned is important only in the heating of air layers directly in contact with the ground.

Convection is the most important means of heat transfer in weather processes. It is the transfer of energy through vertical currents. When, for example, a layer of air close to the ground is heated by the earth, it expands, decreases in density, and, being lighter than surrounding cooler air, rises. This rising current of air is called a thermal, and the process is called thermal convection. Sometimes, air rises when it strikes hills or mountains or other objects in its path; this is called mechanical convection.

Radiation is the emission and propagation of waves transmitting energy with the speed of light through space, with no material medium needed to transmit it. It is radiation that provides the earth with energy from the sun.

What do such terms as overcast, broken sky, and ceiling mean?

These terms, commonly used in weather reports, indicate—on a scale of 0 to 10—the portion of sky covered by clouds.

When there is full cloud coverage and no blue sky can be seen, the cloudiness is $^{10}/_{10}$ and the sky is called overcast.

A clear sky means either no clouds at all or clouds covering less than $^{1}/_{10}$ of the sky.

Scattered means clouds covering up to $^{5}/_{10}$ of the sky.

Broken sky means that more than $^{5}/_{10}$ but not more than $^{9}/_{10}$ of the sky is covered by clouds.

Overcast means a cloud cover of more than $^{9}/_{10}$ of the sky.

Ceiling is the height above ground of the lowest clouds covering more than half of the sky. For example, if there were clouds at 4,000 feet covering half the sky (scattered) and at 7,500 feet there were clouds covering from $^{5}/_{10}$ to $^{9}/_{10}$ of the sky, the ceiling would be 7,500 feet.

Ceiling is unlimited when there are no broken or overcast cloud layers. There can be scattered clouds at several levels, but as long as each level does not cover more than half the sky, there is no ceiling.

Ceiling zero means that the ceiling is 50 feet or less.

How much dust is in the atmosphere?

Even in mid-ocean, air contains 500 to 2,000 microscopic and submicroscopic particles per cubic centimeter, and in dusty cities it contains more than 100,000.

It isn't only man who puts them there. Volcanoes can contribute significantly. For three years after the explosion of the volcano Krakatoa in the East Indies in 1883, brilliant twilight colors were noticed all around the world as the dust spread to encircle the globe.

How big are raindrops and how fast do they fall?

They range from $\frac{1}{100}$ to $\frac{1}{4}$ inch in diameter but usually are less than $\frac{2}{100}$ inch in diameter. They fall at a rate of about 7 miles an hour, or 500 feet a minute.

How much rain falls in a thunderstorm?

Thunderstorm rain duration, of course, varies; average length is about twenty-five minutes. The most intense downpour generally occurs within two to three minutes after the first rain falls, and the precipitation remains heavy for five to fifteen minutes, then decreases (but more slowly than it increased). Tremendous amounts of water can be released. It has been estimated that when a thunderstorm drops about ¾ inch of rain over a 9-square-mile area, something more than half a million tons of water come down.

How much water is in the atmosphere?

If all of it could be condensed, it would make a layer about 2 inches deep in the tropics but only a trace over the polar regions. Worldwide, it would average about 1 inch.

Interestingly enough, worldwide rainfall averages about 35 inches annually—about 35 times the total atmospheric water content—which of course means that all of the moisture is turned over about once every ten days.

How are regions of heavy and light rain classified?

Regions where annual rainfall is less than 10 inches account for about 25 percent of the total land area throughout the world. They are classified as *arid*.

About 30 percent of total land area consists of regions receiving 10 to 20 inches of rainfall annually, and these are called *semiarid*.

Subhumid areas, which receive 20 to 40 inches annually, account for 20 percent of total land area.

Humid areas receiving 40 to 60 inches account for 11 percent of total land area, while other humid areas receiving 60 to 80 inches account for another 9 percent.

The remaining 5 percent of the total land area receives more than 80 inches annually and is classified as *very wet*.

An annual rainfall of between 20 and 100 inches is favorable for agriculture. Semiarid areas are suitable for grazing and dry farming but not for intensive agriculture except under irrigation. Where rainfall is below 10 inches, desert conditions exist.

Mount Waialeale, Hawaii, holds the record as the wettest place on earth, having received an average annual rainfall of 471.68 inches over a period of thirty-seven years, from 1912 to 1949. Not far behind is Cherrapunji, India, with an average annual fall of 450 inches.

At the other extreme, the average annual rainfall is only 1.33 inches in Helwan, Egypt, and 1.56 at Greenland Ranch, California. There are considerable areas in southeastern California, western and southern Nevada, and extreme western Arizona where the rainfall is less than 5 inches a year.

What have been the world's greatest rainfalls?

Here are some heavy ones for various durations:

DURATION	INCHES	LOCATION	DATE
1 minute	1.23	Unionville, MD	7/4/56
15 mins.	7.80	Plumb Point, Jamaica	5/12/16
42 mins.	12.00	Holt, MO	6/22/47
2 hrs., 10 mins.	19.00	Rockport, WV	7/18/89
12 hrs.	52.76	Belouve, La Reunion	2/28–9/64
24 hrs.	73.62	Cilaos, La Reunion	3/15–16/52
2 days	98.42	Cilaos, La Reunion	3/15–17/52
3 days	127.56	Cilaos, La Reunion	3/15–18/52
4 days	137.95	Cilaos, La Reunion	3/14–19/52
5 days	150.73	Cilaos, La Reunion	3/13–18/52
6 days	159.65	Cilaos, La Reunion	3/13–19/52
7 days	160.81	Cilaos, La Reunion	3/12–19/52
8 days	162.59	Cilaos, La Reunion	3/11–19/52
15 days	188.88	Cherrapunji, India	6/24–7/8/31
31 days	366.14	Cherrapunji, India	7/61
6 months	884.03	Cherrapunji, India	Apr.–Sept. 1861
1 year	1041.78	Cherrapunji, India	8/60–7/61
2 years	1605.05	Cherrapunji, India	1860–61

Why can it sometimes rain even though relative humidity is below 100 percent?

Warm moist air can overrun a cooler, drier air mass on the earth's surface, and rain may fall from the warm layers of air through the cooler, drier air below. So even while relative humidity at the surface is reflecting the lesser amount of moisture in the lower layer of the air, rain falls.

When is a snowfall called heavy?

When snow reaches depths of 4 inches or more in a twelve-hour period or 6 inches or more in a twenty-four-hour period.

What have been record snowfalls?

The deepest snowfall in a twenty-four-hour period occurred at Silver Lake, Colorado, on April 14–15, 1921. The accumulation: 75.8 inches.

On January 19, 1933, 60 inches fell on Giant Forest, California. And on January 21, 1935, 52 inches covered Winthrop, Washington.

Perhaps the greatest seasonal snowfall in the United States occurred in the winter of 1955–56 at Paradise Ranger Station in Mount Rainier National Park, Washington, where 1,000.3 inches accumulated.

How much hail can fall in a storm?

It's estimated that a single storm can dump thousands of tons of hailstones. One storm in Chad, Africa, in May 1935 left hailstones piled 19 inches high over a square mile area. In a storm in Washington County, Iowa, in September 1897, there were hailstone drifts as high as 6 feet.

How devastating can hailstorms get?

One that struck Kemptville, a little town south of Ottawa, Ontario, on June 26, 1952, lasted fifteen minutes during which it broke an estimated 5,000 windows, killed 500 dogs, cats and fowl, severely damaged 200 automobiles, punctured 100 metal roofs, and ruined 15 acres of demonstration and experimental crops. Fortunately, few people were injured, none seriously.

One of the worst hailstorms in terms of loss of human life occurred on April 30, 1888, in the Moradabad and Beheri districts of India. Hailstones reported to have been as large as cricket balls killed 246 people, some of them hit and battered by the stones, others buried under great drifts and dying of exposure. More than 1,600 animals were killed.

Still another on June 19, 1932, in Honan Province, China, killed

200 people, injured thousands of others, and destroyed houses, crops and trees over an area containing some 400 villages.

What's the difference between a light and a thick fog?

Actually, fogs are classified in four densities: light fog, with a visibility of ⅝ mile or more; moderate, with visibility of 5⁄16 to ⅝ mile; thick, with visibility of ⅕ to 5⁄16 mile; and dense, with visibility less than ⅕ mile.

Wind speeds are expressed in miles per hour and knots. What's the difference?

In the United States, wind speeds have generally been stated in statute miles per hour, one mile equaling 5,280 feet. More and more, however, wind speeds are being expressed in knots, a term commonly used in marine and aviation work. A knot is equal to one nautical mile per hour, a nautical mile being 6,076 feet.

What's the difference between a backing and a veering wind?

A backing wind is one that changes direction in counterclockwise fashion in the Northern Hemisphere, the opposite way in the Southern. If, for example, a wind starts out blowing from the east, then from the northeast, then from the north, and finally from the northwest, it is backing. A veering wind shifts in the opposite, or clockwise, direction.

What do windward and leeward mean?

Windward means the side facing the direction from which the wind is blowing while leeward means the side in the direction toward which it is blowing.

What's gustiness and what causes it?

Gusts are sudden brief increases in wind speed followed by lulls, or slackening. Gustiness can be caused by winds blowing over irregular terrain, producing eddy currents that are superimposed on the main wind flow. Similar currents can be produced by rising warm air.

What's a squall?

Unlike gusts, which are fleeting, squalls are fairly intense winds that persist for several minutes.

What's air turbulence?

It's a disturbed state of the air with irregular vertical currents. It can be caused by wind flowing over uneven surfaces, thermal currents over areas of different warmth, the forcing of air aloft by hill or mountain, or lifting of warm air by cold air along a cold front.

What color is lightning?

Whitish, although it may seem to have other colors depending upon the background—bluish, for example, against yellowish artificial light.

What is twilight?

A period of incomplete darkness either just after sunset or just before sunrise. It results from the fact that soon after the sun sinks below the horizon or just before it rises, its light rays are reflected from atmospheric gas molecules and impurities that provide a faint luminescence.

What makes stars twinkle?

Small shifting air parcels in the atmosphere have different densities, which affects the curvature of light rays.

*   *   *

What's the worst tornado on record?

It's believed to have been the tri-state tornado of March 18, 1925, which began in southeast Missouri, moved across Illinois and ended in Indiana. Moving at a near-record average speed of 62 miles an hour and lasting 3½ hours, it killed 689 people.

A major reason for the high death toll: People had only a few seconds' warning because the storm-cloud base was close to the ground, the sky very dark, and the dust and debris so massive that the funnel could hardly be seen until it was too late to escape.

It's sometimes said that tornadoes can play freakish "pranks." What are some examples?

The storms have been known to pick up whole trains—locomotives and coaches—right off the tracks, sometimes carrying them some distance. One tornado lifted a locomotive, turned it around in midair and set it down on a parallel track, facing in the opposite direction.

Tornadoes have picked up people and carried them many feet. In El Dorado, Kansas, on June 10, 1958, a tornado reportedly pulled a woman through a window and deposited her 60 feet away—alongside a phonograph record titled "Stormy Weather."

Tornadoes have also been known to pluck blankets and mattresses from beds, leaving the occupants gasping but unharmed.

And one tornado was reported to have lifted a crate of eggs and set it down 500 yards away, without cracking a shell.

What's the worst hurricane disaster on record?

Perhaps the deadliest was the tropical cyclone that struck the Bay of Bengal and the River Ganges delta in India on October 7, 1863. It killed about 300,000 people. It also destroyed 20,000 boats of all kinds and produced a 40-foot high wave that inundated 6,000 acres of islands and lowlands.

In the United States, one of the worst hurricane years was 1955 when eleven storms killed some 1,500 people and destroyed $2 billion worth of property.

How far can a hurricane travel over land?

Hurricanes require warmth and moisture. Over land, they cannot obtain adequate moisture, and by the time they have moved 600 to 800 miles inland, they usually have lost enough force to become ordinary storms.

What are waterspouts?

Tornadoes that occur at sea. The funnel cloud is formed in the same way—by unstable atmospheric conditions usually in association with a thunderstorm. Reaching the water's surface, the tornado picks up spray and becomes a waterspout.

What was the most disastrous flood in history?

In September and October 1887, China's Yellow River (Hwang Ho) overflowed 70-foot-high levies and flooded 50,000 square miles of farmlands, reportedly causing 6 million deaths, inundating 300 villages and leaving 2 million people homeless.

Perhaps the worst flood in the United States—the Johnstown Flood—occurred on May 31, 1889, when heavy rains caused the Little Conemaugh River in Pennsylvania to break through the South Fork Dam, sending a 30- to 40-foot wall of water down the valley onto Johnstown. The exact death toll is unknown but there have been estimates that 2,100 fatalities occurred as the water swept down the valley at speeds of 22 feet a second, razing every tree and building and everything else in its way.

What happened in the famous blizzard of 1888?

On Saturday evening, March 11, 1888, two large storm systems interacted over the East Coast, producing heavy snowfall over an area from Maine to Washington, D.C. After colliding, the two storms stalled for thirty hours.

Connecticut and Massachusetts got 40 to 50 inches of snow. On the ocean, from Nantucket to Chesapeake Bay, 200 ships were sunk, blown ashore or severely damaged. One observer wrote that the storm "drove the seas before it so violently that many tides did not resume their normal heights for nearly a week along coastal ports."

In Washington, which was hit first, wind and ice tore down all telephone lines connecting Grover Cleveland's White House with the Capitol and government agencies. Trains stopped running.

New England towns were buried under drifts as high as rooftops. In New York City, where the storm was most concentrated, 21 inches of snow fell but 70-mile-an-hour winds piled it in drifts more than 20 feet high, marooning thousands of people in horsecars and elevated trains. Those stranded in the trains had to be rescued by free-lance ladder-bearers—at 50 cents a head. Reportedly, twenty Brooklyn mail carriers collapsed in the snow while making their appointed rounds—"a quaint memory," a *New York Times* editorial called it recently. By the time the blizzard was over, 200 New Yorkers had died of exposure.

The weathermen, I am sorry to say, were taken by surprise.

There have, of course, been other severe blizzards. Perhaps one of the worst was the four-day blizzard of February 1966. It caused at least 208 deaths in the eastern part of the country, submerged the northern plains under a blanket of snow, stranded travelers, and killed 96,000 cattle, sheep and hogs in South Dakota alone. Syracuse, New York, got 53 inches of snow; Rochester, New York, 28 inches.

Is there any system of continuous broadcasts of storm warnings and weather observations and forecasts?

Yes, there is, although many people are still unaware of it.

The National Oceanic and Atmospheric Administration (NOAA) has VHF-FM broadcasts that provide continuous weather information and emergency warnings when necessary.

The broadcasts are transmitted on frequencies of 162.55 KHz and 162.40 KHz from National Weather Service offices twenty-four hours a day. The system works this way: During good weather, the latest observations and forecasts are tape-recorded by local National Weather Service offices in messages lasting from three to five minutes. These are replayed continuously so they can be picked up any time of day or night. The messages are revised every three to four hours or, when necessary, more frequently.

When severe weather threatens, forecasters at the local NWS office interrupt the broadcasts with storm warnings, either tape-recorded or live, depending upon the situation.

The 162.55 and 162.40 frequencies require narrow-band FM receivers of 5-kilohertz deviation. Some relatively low-cost AM-FM receivers have a special weather band to receive these frequencies.

The Weather Radio system also has a feature that allows people to be alerted automatically when dangerous weather such as a tornado or flash flood threatens. Radio receivers available for as little as $35 will silently monitor the weather broadcasts and automatically sound a siren or come up to audible volume when the NWS forecaster presses a button in his office to indicate a forthcoming storm bulletin.

Worth noting: The effective range of the broadcasts is about 40 miles, depending upon terrain and receiver quality. Transmission is, like television, by line of sight and may be interfered with by hills, nearby buildings, or commercial radio transmitters. NWS advises buying a receiver only after you are allowed a home trial to make certain it will pick up the broadcasts.

Also worth noting: Receivers are available that can pick up NOAA Weather Radio broadcasts while you travel on interstate highways.

There are now more than 225 Weather Radio stations. When the system is completed, which is expected to be by the time you read this, there will be about 340 stations covering the areas in which 90 percent of the population of the United States lives.

Glossary of Weather Terms

Air mass: A large body of air with nearly uniform horizontal distribution of temperature and moisture.

Anemometer: An instrument for measuring wind speed.

Anticyclone: Also called a *high* or *high-pressure area*. An atmospheric pressure system with relatively high pressure at its center and winds that blow clockwise and outward in the Northern Hemisphere.

Atmospheric Pressure: Also called *barometric pressure*. The weight of the total mass of air above a given point.

Backing: The counterclockwise shifting of winds, for example, from northeast through north to northwest.

Blizzard: Cold, high winds with heavy driving snow.

Clear: Sunny with no clouds or just a few scattered clouds.

Cloudy: Cloud coverage is from three-quarters covered to overcast, with practically no sunshine.

Cloudy, Partly: Approximately one-half of the sky is covered with clouds. Periods of sunshine can therefore be expected.

Cold Wave: A sharp drop to below-freezing temperatures in a short period of time.

Cyclone: An atmospheric pressure system with relatively low pressure at its center and winds blowing inward and counterclockwise in the Northern Hemisphere. Also called a *low,* or *low pressure system*.

Deepening: A pressure decrease at the center of a storm.

Depression: An area of low pressure.

Dew: Water condensed on objects near the ground whose temperatures are above freezing but below the dew point. White dew is frozen dew resulting from temperatures falling below freezing after dew has formed.

Dew Point: The temperature at which air will no longer hold water and precipitation can be expected. If, say, the dew point is 38 degrees Fahrenheit and the temperature is 41 degrees and falling, you can expect some moisture. The greater the spread between temperature and dew point, the less chance of rain.

Discontinuity: A meteorological term for the rapid variation of pressure or temperature at a front.

Drizzle: Very small water droplets (less than 0.5 millimeter diameter) that seem to float in air following air currents and fall at an intensity of less than 1 millimeter per hour.

Eye of the storm: An area, approximately circular, of light winds and fair weather in the center of some tropical storms.

Front: The zone between two different types of air masses.

Front, cold: The zone between advancing cold air and retreating warm air.

Front, warm: The front, or zone, between advancing warm air and retreating cooler air.

Front, occluded: A mix of two fronts occurring when a cold front overtakes a retreating warm front, forcing the warm air up.

Front, polar: A semipermanent discontinuity that separates cold polar easterly winds from warmer westerly winds of the middle latitudes.

Front, stationary: A front that shows little movement over a six-hour period.

Frost: The depositing of ice on ground objects during cold, clear nights.

Gale: A wind speed ranging from 32 to 63 miles per hour.

Glaze: A coating of clear smooth ice on exposed objects caused by freezing of a film of supercooled water deposited by rain, drizzle, fog.

Hail: Balls or lumps of ice, often with concentric layers of clear and milky ice, greater than 5 millimeters in diameter. If the diameter is less, called *ice pellets, snow pellets,* or *small hail.*

High: A pressure system with relatively high pressure at the center and winds that blow clockwise and outward in the Northern Hemisphere and counterclockwise and inward in the Southern Hemisphere.

Hygrometer: An instrument for measuring humidity.

Instability: An atmospheric condition in which air, after an impulse to move vertically, tends to continue to move farther from its original level.

Inversion: An atmospheric condition in which temperature increases with increasing altitude instead of decreasing as it normally does.

Jet stream: A narrow stream of strong winds in the upper atmosphere.

Land breeze: A light wind that blows from land to sea at night after the land has become cooler than the sea surface.

Low: A pressure system with relatively low pressure at its center and winds that blow inward and counterclockwise in the Northern Hemisphere and outward and clockwise in the Southern Hemisphere.

Mackerel sky: A name for cirrocumulus or altocumulus clouds.

Millibar: A unit of pressure equal to 0.295 inches of mercury.

Mountain breeze: A wind that blows downslope along mountainsides into adjoining valleys, most commonly at night as air in contact with mountain slopes cools.

Ozone layer: An upper atmospheric layer with a concentration of ozone, a form of oxygen containing three atoms, instead of two, to the molecule.

Precipitation: Falling particles such as rain, snow, hail or drizzle.

Radiosonde: A balloon-borne instrument for simultaneously measuring and transmitting meteorological data.

Rain: Droplets greater than 0.5 millimeters in diameter, with an intensity generally more than 1.25 millimeters per hour. The droplets, while larger than those of drizzle, are fewer, and there is generally less reduction in visibility except in the case of heavier rainfalls.

Rain, freezing: Rain that freezes into icy sheets on the ground and on trees, wires, etc.

Rain, occasional: Rain that falls intermittently.

Relative humidity: The ratio of the amount of water vapor in air to the amount that would be there if the air at the same temperature were saturated with vapor.

Ridge: An elongated area of high barometric pressure.

Rime: White or milky granular deposit of ice formed by rapid freezing of supercooled water droplets hitting an exposed object.

Saturation: A condition in which air holds a maximum amount of moisture; 100 percent relative humidity.

Scattered showers (or **thunderstorms**): Showers occurring in one vicinity and not another. A local shower may soak one side of the street and not the other. If you're lucky, you can have sunshine all day with such a forecast.

Sea breeze: A usually light wind blowing from sea to shore when the land is warmer than the sea surface.

Secondary cold front: Another cold front that may develop behind the first and may bring even colder air with it.

Secondary depression: A low-pressure area forming to the south or east of a primary depression.

Semipermanent high or low: One of several relatively stable and stationary atmospheric pressure and wind systems, such as the Bermuda high or the Icelandic low.

Sleet: Transparent solid grains of ice formed from freezing of rain or slight melting and refreezing of snow as it falls.

Small-craft warning: A warning posted for small boats by coastal weather stations when winds with velocities of more than 25 miles per hour are expected.

Snow: White or translucent ice crystals often agglomerated into the form of flakes.

Snow grains: Very small white opaque particles that do not shatter or bounce when they hit a surface and are the solid equivalent of drizzle.

Snow pellets (soft hail): White, opaque, round or conical ice particles with snowlike structure, 2 to 5 millimeters in diameter, crisp, easily crunched.

Source region: An area over which an air mass acquires its more or less uniform temperature and humidity horizontal distribution.

Squall: A strong wind, usually of 16 knots or more, sudden but continuing for at least two minutes, distinguishing it from a gust.

Squall line: A line of instability in advance of a cold front, with strong winds, turbulence and often heavy showers.

Stability: An atmospheric condition in which air that has moved away from its original level will tend to return to that level.

Synoptic: Pertaining to or affording an overall view. In meteorology, use of data obtained simultaneously over a wide area for presenting a comprehensive and nearly instantaneous picture of the state of the atmosphere.

Tropical cyclone: An area of low pressure originating in the tropics.

Trough: An elongated low-pressure area.

Veering: The clockwise shifting of winds, for example, from northeast through east to southeast.